# Theaters of Anatomy

# Theaters of Anatomy

*Students, Teachers, and*
*Traditions of Dissection in Renaissance Venice*

CYNTHIA KLESTINEC

The Johns Hopkins University Press
*Baltimore*

© 2011 The Johns Hopkins University Press
All rights reserved. Published 2011
Printed in the United States of America on acid-free paper
2   4   6   8   9   7   5   3   1

The Johns Hopkins University Press
2715 North Charles Street
Baltimore, Maryland 21218-4363
www.press.jhu.edu

*Library of Congress Cataloging-in-Publication Data*
Klestinec, Cynthia, author.
Theaters of anatomy : students, teachers, and traditions of dissection
in Renaissance Venice / Cynthia Klestinec.
p. ; cm.
Includes bibliographical references and index.
ISBN-13: 978-1-4214-0142-3 (hardcover : alk. paper)
ISBN-10: 1-4214-0142-8 (hardcover : alk. paper)
1. Human dissection—Italy—History—17th century.   I. Title.
[DNLM:   1. Anatomy—history—Italy.   2. Anatomy—education—
Italy.   3. Dissection—education—Italy.   4. Dissection—history—
Italy.   5. History, 16th Century—Italy.   6. History, 17th Century—
Italy. QS 11 GI8]
QM33.4.K64 2011
611—dc22      2010049755

A catalog record for this book is available from the British Library.

*Special discounts are available for bulk purchases of this book. For more
information, please contact Special Sales at 410-516-6936 or
specialsales@press.jhu.edu.*

The Johns Hopkins University Press uses environmentally friendly
book materials, including recycled text paper that is composed of
at least 30 percent post-consumer waste, whenever possible.

*For Andrew and Stella*

# Contents

# Figures

# Preface

This book tells a new story about the history of anatomy, one concerned with the role of the anatomical theater, with the relationship between teachers and students, and with the differences between public and private anatomical instruction. These features reveal the conflicted and contested importance of *seeing* anatomy, compelling us to examine and revise our assumptions about the relationships between anatomy, observation, and early science. They also explain how Padua established itself as a center for anatomical research in the period between Andreas Vesalius and William Harvey.

Drawing from traditional sources, such as anatomy texts and institutional documents, and new sources, such as the letters and records of students, *Theaters of Anatomy* invites the reader to reconsider the post-Vesalian era of anatomical inquiry. The post-Vesalian era covers the development of anatomical inquiry in the wake of Vesalius's departure from the university of Padua in 1542.[1] In this era, the Venetian-controlled city of Padua is particularly remarkable because it hosted a succession of famous anatomists—Realdo Columbo (1516–1559) and Gabriele Falloppio (1523–1562), as well as Girolamo Fabrici (1533–1619), Giulio Casseri (1552–1616), Adriano Spigelio (1578–1625), Johann Veslingus (1598–1649), and, as a student, William Harvey (1578–1657). Among these figures, Fabrici deserves special attention because he instituted a natural philosophical program of anatomical inquiry and held the chair in anatomy and surgery from 1565 to 1609, a long period in which he became a powerful figure at the university.[2] Rather than substitute Fabrici for Vesalius in a familiar story of progress, however, I use Fabrici (and his successors) and his students to bring into focus the several forces that interacted and came to constitute the public and private worlds of university anatomy between the 1550s and the 1620s.

Fabrici's program was intimately linked to the construction of two permanent and increasingly public anatomical theaters, built consecutively in the central building of the university in 1584–1585 and 1594–1595. Both theaters were a sign of the strong, supportive relationship between the university and the Republic of Venice and a portent of anatomy's growing academic and cultural significance. In these new places, a highly dramatic tradition of public anatomy demonstrations emerged; with music in the background, anatomists engaged in the study of anatomy and promoted it for a varied audience of professors and students, as well as shoemakers, fishmongers, and basket weavers. The demonstration evolved from a purely academic event into a spectacle rich with symbolic and civic significance. It publicized the study of anatomy and quickly transformed the demonstration into an event with wider cultural significance. Anatomists enjoyed frequent pay raises and, for a time, increasing numbers of students. Finally, while it and other centers of medical education were well known for their anatomical publications, Padua was additionally known for its anatomical theaters and its anatomical performances.[3]

In tandem with this public tradition, a private tradition of anatomical lessons and dissections survived and, in the final decades of the century, flourished. The private tradition acquired greater significance in the post-Vesalian period because the public tradition of anatomy consolidated itself so fully around the philosophical tenets of Fabrici's program. The private venues of anatomy came to be seen, by students especially, as contrasts to the public arena. If the public arena elaborated the philosophical dimensions of anatomy and more contemplative inquiries about nature, the private arenas brought students face to face with cadavers, with the procedures of dissection and vivisection, and with expressions of technical, manual skill. In private, students witnessed a form of expertise, embodied in the anatomist, that depended on manipulating the cadaver, carefully distinguishing its parts, and expanding the study of anatomy with inquiries into morphology (the relationship between structure and function) and physiology (the study of coordinated processes). In one instance, an anatomist set up an elaborate surgical device to demonstrate the movement of the muscles, shuttling his small audience from corpse to instrument in an analogical tour-de-force. These novelties emerged in the private realm because of the ways in which the public and private traditions of anatomy had evolved to host presentations that contrasted in style as well as in content.

Despite such intriguing changes, the post-Vesalian period has not been fully characterized. Its developments have been kept distant from the main lines of thought regarding the history of anatomy. The period has been treated as a kind of distraction from Vesalius or a dilution of his message—to look and see for yourself—on the one hand, and an unnecessary delay before the appearance of Harvey and his discovery of blood circulation (1628), on the other. In other words, this period has been seen as provisional because Vesalius's approach to the study of anatomy is taken as the nearly direct line of inspiration for Harvey's experimental proclivities.[4] While new studies on Vesalius have emphasized his humanist aspirations and his commitment to Galen and the textual traditions of anatomy, the assumption that Renaissance anatomy merely introduced a new kind of visuality or a new visual economy for the study of nature remains and, with it, a diminished significance for this period.[5] Certainly new ideas about seeing, observation, and visually derived evidence emerged in the early modern traditions of anatomy, but it is only by investigating this provisional period that we can understand what they were and how and why they came about in the shapes that they did. This book thus draws attention to various features of anatomical inquiry, not just those that depend on sight. It looks at the cultivation of anatomical learning among students, both in decorated anatomical theaters and in private lessons, emphasizing the pedagogical, institutional, and rhetorical contexts of anatomical inquiry alongside the experiences of medical students.

Each chapter examines the relationship between the public and private traditions of anatomy and the activities, concerns, and experiences of medical students. Each chapter also emphasizes the pedagogical and rhetorical aspects involved in the presentation of anatomy. The public-private distinction is explored in the context of the pedagogical habits of Falloppio in the 1550s and 1560s (chapter 1). Fabrici distinguished himself from Falloppio, but the former's pedagogical innovation led to students' complaints, institutional reform, and a good deal of academic rivalry, setting the stage for the first permanent anatomical theater (chapter 2). The second anatomical theater brought greater public awareness and recognition to the study of anatomy and greater transparency to its procedures by hosting dramatic anatomies, spectacles that emphasized natural philosophy and trained students to be more obedient and silent spectators (chapter 3). This tradition shifted the wider community's suspicions about dissection from the public to the private realm and from the professor to the students (chapter 4). The private venues of anatomy were open to novel combinations of anatomical and surgical knowledge, which fostered an appreciation of manual expertise and habits of visual scrutiny (chapter 5). The changes in anatomical inquiry in the post-Vesalian era reveal an energetic, visually attuned tradition of private inquiry, one that influenced William Harvey's approaches to anatomy. Equally, these changes attest to the embedded nature of early scientific practices and to the necessity of characterizing the spaces of inquiry to the fullest degree possible.

To explore the role of anatomical theaters in the pedagogical culture of the university, this book is organized chronologically around the public and private traditions of anatomical inquiry and is focused, in particular, on the impact of the second anatomical theater. Where possible, the records of medical students are placed foremost in the analysis. These students remind us that the practices associated with observation often migrate to the center of early scientific communities in unexpected ways: through alternative pedagogies, student participation, and the informalities governing academic relationships.

Chapter 1 describes the structural origins of anatomical spectacles by charting the history of the public demonstration of anatomy alongside private anatomical exercises. Early on, the public demonstration served as an introduction for students; there, students learned the general outlines of human anatomy as well as the importance of understanding anatomy for its natural philosophical purpose. They also learned that with anatomical knowledge, they would be able to practice medicine more effectively because they could isolate the causes of disease in the body and perform surgical procedures such as bloodletting, cauterizing, and setting fractures (procedures considered essential in the education of both physicians and surgeons). These features were introduced in the public demonstration, alongside current medical debates—that is, in the commentary tradition of anatomy—but they were studied further in private anatomical lessons. In private, students could gain a better sense of the practical dimensions of anatomical knowledge as well as an increased visual and tactile appreciation of the human body.

Following this early history of public and private anatomies and the centrality of pedagogy in Renaissance anatomy, chapter 2 shows that Fabrici, who succeeded Falloppio, chose to develop more fully the study of anatomy as it related to natural philosophy. This was due in part to a perceived crisis in anatomical pedagogy, and it had several consequences. It downgraded both the importance of discovering new structures in the body and the procedures of cutting open corpses. The language of discovery was replaced by a rhetoric of self-revelation: the cadaver was said to reveal its interior structures, while the anatomist would speak about its formal and philosophical significance. The work of dissection was gradually but profoundly superseded by natural philosophical reflection. This new style of instruction elicited severe criticism from medical students; they complained loudly and often. They eventually sought to reform the curriculum so that anatomy lessons included introductory-level instruction, a strong focus on the structural or morphological aspects of anatomy, and opportunities to learn the procedures of dissection and surgery. Students requested—and then demanded—more private lessons in anatomy and different teachers. Into the fray came the first permanent anatomical theater (1583–1584), and it solidified the link between professors and the spaces where learning occurred. Fabrici taught in public and in the theater, while other anatomists taught in private and in other, smaller locations. That link became the main axis for the rivalries between professors and for the separation of anatomy as a natural philosophical inquiry, on the one hand, and anatomy as a necessity for the practice of medicine and surgery, on the other. A public demonstration reflected a stronger philosophical orientation, while a private lesson developed a clearer focus on procedures for cutting (applicable to dissection and surgery).

Until the 1580s, anatomists tended to float between these two venues. This allowed content to track more consistently between public and private exercises. In the 1580s, however, Fabrici began to use the public event as an occasion to present his research. In other words, he narrowed the topics that he would treat, making the public event into a specialized venue; this was a new development in the history of anatomical presentations. He also sought to be the only professor allowed to host these public events, which left his colleagues to conduct their work during private lessons. The division of labor was a new one. Before the 1580s, a public anatomy demonstration provided an introduction to general anatomy and to a few of the current debates in the field of medicine; after the 1580s, it evolved into highly topical, focused discussions on the natural philosophical dimensions of anatomy. At these public demonstrations, the audience was too large to be able to see the partially decomposed particulars that were revealed in the corpse. The processes of dissection became a part of the preparation for the demonstration, not a part of the demonstration itself. A private dissection allowed students to see the parts of the cadaver or the animal and the techniques used to dissect them; a public demonstration did not. Unlike private dissections, the public demonstrations were used to extend the relationship between natural philosophy and anatomy; they relied on the verbal skills of the anatomist and the listening skills of his audience.

Examined within the context of public and private traditions of anatomy, the construction of permanent anatomical theaters highlights the support for and refinement of the public tradition of anatomy. As chapter 3 shows, the second permanent anatomical theater, built in 1594–1595, further solidified the boundary between public and private anatomies and amplified the rivalries between professors. For example, the second theater has Fabrici's name engraved above the entrance, signaling that the theater and the public anatomy demonstration belonged, in a certain sense, to him.

Because Fabrici organized the demonstrations around his research interests, specific topics, and units of anatomy, the public anatomy demonstration did not focus on dissection per se. Nor did its format derive from the progressive dissection of an entire human body. Although the dissected body was fundamental to Falloppio's and other anatomists' demonstrations—it was the very product of the demonstration—it declined in significance by the end of the sixteenth century. Fabrici's philosophical orientation combined with the formal atmosphere and aesthetic features of the new theater to create anatomy demonstrations that relied on philosophical explanations and music for their structure, rather than on the progressive stages of human dissection. In the theater and before Fabrici, students quietly contemplated these aspects of nature, learning the philosophical dimensions of anatomy more by listening than by seeing.

Chapter 3 also underscores the role that the public anatomical theater played in shaping the students' behavior, that is, their comportment.[6] It disciplined the bodies and minds of students by limiting disruptions (fueled by the intense rivalries and escalating violence between local and foreign students) as well as the potential for formal debate (disputation). As disputation waned and contemplation was given a new emphasis, students recognized the importance of silence. Their testimony reveals the strange collusion between silence, contemplation, and restraint, a triad that suggests a new disposition toward anatomical inquiry, one clearly evident in Harvey's works.[7] In Padua, medical students had at their fingertips the comportment manuals of Giovanni della Casa, Stefano Guazzo, Bartolomeo Meduna, and others, authors who placed a premium on a well-behaved, physically and mentally disciplined student. While such postures were surely ideal—more prescriptive rather than descriptive—they would be evoked and reinforced by the rhetorical conceits of late sixteenth-century anatomy, by the call to contemplate anatomy, and by the protocols instituted in the anatomical theater. In this sense, anatomy became a more integral part of humanist education; its practices reveal forms of discipline and internal regulation that the students themselves began to articulate.

While literary scholars have described anatomical theaters as "darkly erotic" and historians have characterized them as carnivalesque celebrations of the flesh and of human generation, chapter 3 shows that late in the sixteenth century, the focus on dissection and observation gave way to philosophical inquiry and contemplation. The theater brought a new visibility to public anatomy demonstrations, but it did not entirely ameliorate the persistent connection between the study of anatomy and the disruption of burial rituals. As chapter 4 reveals, the belief that students robbed graves for bones as well as bodies continued, but such fears were localized, in two senses. They were tied to concerns about local rather than foreign corpses and to private anatomical exercises rather than public ones. Working from an account of a sensational crime—a local fruit seller killed his wife by cutting her body into pieces—this chapter suggests that the association between medical students and their dissecting activities became more evident, and thus formed the basis for the wider community's concerns about anatomical inquiry at the end of the sixteenth century. It also shows how medical students handled their own fears regarding dissection, how they learned to undertake a search for multiple cadavers, and how, in doing so, they learned to identify with the academic community and its broad-based support for the study of anatomy.

Chapter 5 turns to the private traditions of anatomy and the criteria developed for the appreciation of manual, technical skill. As the public demonstration evolved into an event that combined philosophy, rhetorical appeal, drama, and spectacle, the private dissection assumed the role of training students in dissecting procedures and surgical maneuvers. Singling out the notable features of private anatomies, students praised anatomists for their ability to dissect beautifully and demonstrate clearly how to manipulate body parts to perform surgical manipulations, such as setting fractures. While forms of manual or technical expertise (*peritia*) might suggest the world of artisans and craftsmen, the inclusion of these skills in private anatomies suggests a link to the world of learned surgeons, especially those practicing and publishing in nearby Venice.[8] By the end of the sixteenth century, and because of the popularity of the anatomical theater, students began to strengthen their commitment to the anatomical education they received in private—it was exclusive in a way that the public anatomy demonstration and the public anatomical theater were not. Amid concerns about the nobility of their education and with an eye to their own postgraduate careers in Venice and beyond, students helped to augment the significance of private anatomies. These developments characterized the university as an exclusive place, but one remarkably open to the necessities and contingencies of the medical profession.[9]

Moreover, these developments introduced a realm of inquiry that was private and safe from the contaminating presence of craftsmen. In the early seventeenth century, Giulio Casseri, the teacher assigned to these private venues, combined pedagogical features from the public and private traditions of anatomy, calling for students to contemplate anatomy, as they had in public demonstrations, and to inspect anatomy, as they routinely did in private ones. With Casseri, the pedagogical merits of this attenuated private venue were clarified: the private venue permitted greater proximity between the anatomist, his students, and the cadavers and animals. In this setting, students witnessed technical expertise, attributed beauty to the dissected corpse, and sustained their visual appreciation of the dissected material. It was here that students learned to combine contemplation with inspection, to put away charts and diagrams and focus on seeing the corpse. Finally, a brief epilogue addresses the impact of these developments on the experimental scenes of Harvey's work and calls attention to the significance of space in characterizing the embedded nature of early scientific inquiry.

My account of post-Vesalian anatomy and its theaters in Padua is a story not only about the fascination with dissected bodies, but also about the alignments between students and officials, and between universities and their local settings. It traces the publicity and visibility of anatomical inquiry: how it was viewed by students and other members of the academic community as well as by the wider, more diverse community that lived beyond the university's walls. It acknowledges that Fabrici's study of Aristotle and his interest in natural philosophy were intellectually significant, but it extends that point by examining the actual features of Fabrici's demonstrations. In his public demonstrations, it was not seeing the dissection of an anatomical part or touching that part that was important, but rather hearing and conceptualizing his new program of anatomical inquiry. This is the context for understanding why and how anatomy demonstrations began to take on dramatic properties and include both musicians and nonacademic spectators. In other words, public anatomies drew spectators from outside the academy when Fabrici was the leading anatomist, when he publicized his philosophical inquiry into anatomy, and when his demonstrations took place within the second permanent anatomical theater. The impact of this development is equally intriguing, because it allowed the private venues of anatomy to emerge as the basis for a coherent, cogent, visually oriented tradition of anatomical inquiry. That tradition should be understood as the true legacy of the post-Vesalian era.

Although the Venetian year began on March 1, I have followed current scholarly fashion and modernized the calendar. Unless otherwise stated, consecutive dates, for example, 1588–1589, indicate the academic year rather than two calendar years.

# Theaters of Anatomy

# Introduction

*Redefining the Post-Vesalian Era*

On 12 December 1556, the young anatomist Gabriele Falloppio (1523–1562) ended a letter to the administrators at the University of Padua, with a clear, though indirect, threat. Despite their "immense ardor" for the topic, Falloppio explained, medical students had not had an anatomy demonstration in two years and, with the students not finding preparations for one underway, "they begin to make plans to go to Bologna or to Ferrara, where undoubtedly they will have these spectacles [*feste*]."[1] In an age of anatomical and other revolutions, the idea that actual dissections were not taking place with frequency or regularity is not so much a surprise as a disclosure akin to a dirty secret. Just as other disciplines associated with the Scientific Revolution focused their material practices on the stuff of nature, so the historical changes in anatomy had to depend, at the very least, on more frequent dissections. Indeed, Falloppio succeeded the well-known anatomist Andreas Vesalius (1514–1564), whose zeal for dissection is often cited as a key source for the widespread enthusiasm for anatomy in the Renaissance.[2] Falloppio's admonition to the administrators, however, alerts us to the considerable irregularity of anatomical exercises and the extent to which students, working closely with their professors, were essential to the cultivation of anatomical knowledge in the period. It was the students' desire that motivated Falloppio to inquire about the annual anatomy demonstration; and it was surely the threat of their departure that motivated administrators and the institution to grant Falloppio's request to host such spectacles.

Despite their evidently important role, students have not figured prominently in historical studies of anatomy and dissection, even when those histories emphasize the practical or material dimensions of the subject.[3] Their absence has obscured a variety of features related to the ways in which anatomical knowledge was presented and learned in the early modern period. These include the significance of both temporary and permanent anatomical theaters, the sometimes contested relationship between students and teachers, and the role of public and private instruction. Silence around these issues, like the obscure role that students have played in histories of anatomy, is part of a lingering gap between theory and practice in the historiography of the Scientific Revolution. While this gap has its methodological roots in the traditions of intellectual and social history, I aim to bridge it with an analysis of the anatomical spectacle as it shifted from temporary to permanent theaters, as it engaged students and teachers, and as it was conducted in public and private venues. Thus, this introduction will elaborate these three areas of concern in order to begin an inquiry into the anatomical spectacle, which rose to such prominence in Padua in the second half of the sixteenth century—in the period between Vesalius and Harvey—that it earned the university a reputation as an innovative institution.

Inside anatomical theaters and in both public and private demonstrations, students and teachers engaged in the study of anatomy, but questions remain about the details of their study. Students and teachers worked within intellectual frameworks, drawn from Galenic medicine and Aristotelian natural philosophy, and with material specimens. They sought to understand the parts of cadavers through the terms and categories that came from their medical and philosophical readings. How did students learn to do this? And what consequences do these learning practices have for our understanding of the post-Vesalian era of anatomy? Students were not passive receptacles for the ideas of their teachers. Instead, they responded critically, sometimes privileging one kind of knowledge over another and sometimes acknowledging that a certain explanation was insufficient. When a philosophical explanation felt inadequate, for example, it could then redirect students to a scrutiny of the corpse and draw attention to the manual skills needed for dissection. Such skills tended to be associated with barbers, surgeons, and sometimes butchers, but students helped to transform manual skill into an achievement and a mark of expertise, a laudable skill rather than a liability. These changes took place in part because the anatomical spectacle elevated aspects of the anatomical tradition, such as its technical capacity, into a praiseworthy attribute. The anatomical spectacle, moreover, became the occasion for professors to think and write about the hard work needed to develop habits of attention and concentration in their students (traits familiar to us as precursors to the Enlightenment and modern science); set within the subject of anatomy, students became the objects of these discussions and measures.[4] The figure of the student allowed these ideas to coalesce into habits inculcated in him, but just how malleable was the Renaissance medical student? The written accounts left by medical students provide answers to these questions, often in breathtaking detail, and, when combined with anatomy texts and institutional documents, they offer a more complete and compelling picture of the philosophical and social dimensions of the history of anatomy, circa 1550–1620.[5]

Against a background of the work of their teachers and the culture of the university, students come forward to describe the nature of their experiences with anatomy. They direct attention to the role of manual skill and dissecting techniques, to the philosophical dimensions of anatomical knowledge, and to the professors, corpses, and animals they encountered. They respond critically to their professors, revealing a nuanced appreciation of the success and failure of particular anatomical lessons. They also praise their peers for demonstrating a broad knowledge of the textual traditions of medicine, including herbals and anatomy, and for acquiring technical expertise in dissection and surgery. These new sources of information offer a much more complicated picture of the spectacles Falloppio mentioned in his letter. According to these student records, anatomical inquiry did not depend on gory spectacles of dissection or solely on the visual scrutiny of internal and external body parts. Rather, the development of these "spectacular anatomies" depended on the rhetorical habits of professors; a pedagogical emphasis on natural philosophy; curricular reforms, especially those initiated by students; and the construction of permanent anatomical theaters. These features were not secondary to a technical and practical tradition of dissection or somehow distant from the theoretical basis of anatomical inquiry; they were fundamental. They transformed the work of the hands into a sign of expertise, and they helped both anatomists and students see the dissected specimen and apprehend its significance in new ways.

## THE ANATOMICAL THEATER

In Padua, the post-Vesalian era was a vibrant one because of its many gifted anatomists and its exquisite anatomical theaters. In the historiography on anatomy and early modern culture, the anatomical theater is described and understood as a place for seeing the dissection of a corpse. By examining illustrations of dissected corpses, literary works, and descriptions of anatomical theaters, Jonathan Sawday, for example, focuses on the "culture of dissection" and argues that Vesalius's work revealed and promoted the idea of visual evidence—derived from the dissected corpse—and that his publication and the public dissections in London's anatomical theater (designed by Inigo Jones in 1636) helped to ground the wider culture's preoccupation with the ultimate mystery of *seeing* the body's interior.[6] More recently, Richard Sugg has attributed the excitement generated by anatomy to "the uneasy thrill of strangeness still provoked by the sight of the human interior" and calls the Renaissance variant of this thrill "a more potent novelty" derived from an emphasis on the dissected body in anatomical theaters rather than from the books that discussed it.[7] The scenes described in early hospitals and studied by Piero Camporesi and, recently, by Jon Henderson suggest that bodily innards and the wounds revealing them were neither thrilling nor all that scarce.[8] Nevertheless, the anatomical theater has remained a key site for this thrill. In his study of the history of learned anatomy, Rafael Mandressi has taken a more measured approach, characterizing the visual emphasis of early modern anatomy not only as an oppositional response to book knowledge, but also as a feature connected to the introduction of the anatomical theater, the place "where one sees."[9] Similarly, in Paula Findlen's essay on the sites for anatomy, botany, and natural history, the anatomical theater is noted for its ability to train visual perception.[10] Such accounts reveal the pervasive assumption that the anatomical theater provided a visual experience, a place to see.

It is difficult to tease apart the visual dimensions of anatomical knowledge and the sight-centered orientation of its practices. The study of anatomy, linked obliquely to the study of perspective,[11] appears to be similar to other visually stimulating bodily encounters of the early modern period: to the semiological trends in therapeutics, where a diagnosis depended on seeing and identifying symptoms and signs of disease; to the corporeal signs of sanctity found in the bodies of female visionaries; and to the imperialist fantasies underwriting early cartography.[12] In addition to its connections with these, anatomical inquiry, in particular, is organized around the sensed perception of structures, around habits of observation and scrutiny, and around the magnetic attraction of an opened corpse. The anatomical theater has served as an emblem for these complex activities, guiding our understanding of how and why they began to dominate anatomical education in the early modern period. Nonetheless, while many studies have focused on the printed images in anatomy texts as clear indicators of its visual appeal, my intent in this book is to query what was actually being presented inside the anatomical theater. Dissection, as a set of procedures and as a staged process of revelation, was often mostly completed (eventually by students) before the public demonstration began. In other words, the spectacular nature of anatomy and the content of the anatomy demonstration did not depend on the gradual process of opening a corpse. Given the historiographical tradition laid out above, this is a real problem. If the process of dissection and the revelation of anatomical structure did not form the basis of the anatomy demonstration (or the performance, if you will), then what was going on in the theater? In what sense was the theater a place "where one sees"?

The spectacular nature of these anatomies depended on the anatomist's ability to present anatomy as a natural philosophical endeavor, that is, to set the material aspects of the decaying corpse or animal within the natural and spiritual orders. He might treat the subject of human anatomy generally or focus on particular topics, but the demonstration involved his verbal presentation of features and problems from the commentary tradition and his own writings, rather than from the abnormal or pathological features present in the corpse or the manual activities associated with dissection. As the students repeatedly imply, these demonstrations were auditory—not visual—events, organized around the silent presence of the corpse and the loquacious presence of the anatomist. These details make it difficult to understand how the public anatomical theater enhanced the visual apprehension of anatomy, how it supported training in visual perception, or, indeed, how it relates to the most basic ideas we have about the history of anatomy. Once these details are aligned with private lessons on anatomy, however, they help us to understand how both venues trained the senses—hearing, sight, and touch—and connected anatomy to the broader traditions of learned medicine and natural philosophical inquiry.

Given the historiographical emphasis on the visual, it may seem counterintuitive to claim the presence or persistence of *nonvisual* components in anatomical study. Yet many of the practices associated with the post-Vesalian era of anatomy do not reflect an overtly visual emphasis. Final exams for medical students, for example, continued to be based on the interpretation of texts rather than on on-site diagnoses. Moreover, the verbal fireworks of public disputations still dominated academic training. Much at the Renaissance university depended on auditory rather than visual perception. This was true even for the study of anatomy, and it was especially true for the studies of anatomy that Girolamo Fabrici of Aquapendente (Hieronymus Fabricius ab Aquapendente, 1533–1619) made inside the anatomical theater, where (it must now be said) much more was in play than the visual elucidation of the components of a corpse. Inside this theater, sight was constantly problematized. Not only did large crowds obfuscate the particulars of the specimen but, more than that, the messy, laborious work of dissection was mostly completed during the preparatory stages of the demonstration, that is, before the crowd assembled and the performance began. This book queries the reasons why medical students and other spectators came to the anatomical theater, if not for the chance to see a corpse slowly opened and its interior compartments dissected.

Based on the perceived connection between the anatomical theater and sight, previous studies have used the visceral experience of watching the dissection of a corpse as a way to explore and characterize the psychological and sociological impact of public dissection on the academic and the wider communities.[13] For these studies, the anatomy demonstration emphasized the corporeal and even the ghoulish aspects of dissection.[14] From such material origins, the aesthetic dimensions of anatomy have been characterized both as part of the tradition of *memento mori* and its themes of death and human mortality, and as part of the tradition of Carnival and its theme of regeneration. Influenced by the work of Mikhail Bakhtin and Michel Foucault, these studies view the function of public anatomies variously: as sites of tragic reflection on the nature of human

mortality; as sites of corporeal punishment extending from the executioner to the anatomist; or as sites of disruption, raucous outburst, and
bawdy display, which reveal the potentially subversive reactions to death
and violence that dissection would seem to elicit.[15] My evidence suggests
that medical students contemplated questions of human mortality in the
course of participating in the annual anatomy demonstration, but this
evidence depends exclusively on the students' participation in the burial
ceremony that typically followed directly on the heels of the demonstration. Students ruminated on the subjects of death and the afterlife at the
time of the burial, rather than during the anatomy demonstration per se.

In late sixteenth-century Padua and inside its acclaimed (second) anatomical theater, the significance of the corpse decreased in the public
events. The dissection was conducted in a smaller, more private chamber, before the demonstration began. The demonstration also featured a
more prominent philosophical component that emerged from Aristotle's
works and the tradition of natural philosophy. These aspects provide a
vivid contrast to the picture of riotous, carnivalesque public anatomies.
This is also informative. Scholars have derived the relationship between
public anatomies and Carnival from the particular case of Bologna and
the anatomical practices and procedures at its university. In her classic
study of Bologna's anatomical theater, Giovanna Ferrari explained that
by the 1640s, the annual anatomy demonstration and Carnival overlapped:
both occurred in the winter months and, more importantly, spectators
came to the anatomy theater wearing carnival masks.[16] This conflation,
Ferrari goes on to explain, was intentional: university officials were aware
that fewer foreign students were coming to the university—matriculation
levels were in decline (severely so by 1640)—and to advertise both its
new anatomical theater (ca. 1638) and the institutional innovation it her-
alded, the administration promoted the association between the annual
public anatomy demonstration and Carnival.[17] Though her account is
particular to Bologna, it has served as a template for understanding the
relationship between theatrical spectacle and anatomy—a relationship
based on the opening of a corpse—and for understanding the historical significance (despite geographical and temporal variation) of the anatomical theaters in London, Leiden, Uppsala, Padua, Bologna, Pisa, those
throughout Spain, and elsewhere.[18]

While more subtle explanations for Padua's differences will appear
in the course of this book, there are two main reasons why the Bologna

model fails to account for the events and traditions in Padua. First, Fabrici held the chair in anatomy in Padua for over fifty years, a tenure that provided a great deal of stability and coherence to the practices of anatomy. That stability and coherence meant that Fabrici could make changes while keeping his own reputation and agenda in mind. When he contacted the Venetian Senate in 1595 to request that the annual demonstration and entrance into the anatomical theater be free and open to the public, he meant to widen *his* audience and to augment his reputation with the noble veneer of privilege, not to combine his demonstration with the ritualized and popular festivities of Carnival. Fabrici still maintained his control over the space, not permitting it to be used for alternative instruction or other kinds of events. In contrast, the theaters in the north—in Leiden (1597), Delft (1614), and Amsterdam (1619)—served as sites of instruction and, according to Jan Rupp, as cultural centers with libraries and museums of curiosities and works of art.[19]

The second reason that a different model is needed for the situation in Padua concerns institutional history and Counter-Reform movements. By the 1640s, the university in Bologna was suffering from a lack of students. In the post-Tridentine environment, the activities of foreign students raised suspicions; as a result, fewer transalpine students were traveling over the Alps to attend Italian universities, particularly those institutions in Catholic regions. In this environment, Italian universities found themselves competing for foreign students by advertising their innovation. The Bologna administration decided to promote their newly constructed anatomical theater not only to advertise its architecture, but also to sell their annual anatomy event by associating it with the festivities of Carnival. Padua, however, was a part of the Venetian Republic, which engineered a different (or at least delayed) Counter-Reform atmosphere. Foreign students continued to attend its university; matriculation did not begin to decline until the second decade of the seventeenth century, long after the permanent anatomical theaters were built and in regular use.[20] At the University of Padua, foreign students were an important, recognized part of the student body. For example, in a decree on voting privileges (ca. 1591), the Venetian Senate called attention to the good behavior of foreign students, the "great valor" of the foreign student body, and the ability for both to enhance the reputation of the university abroad, in *il mondo tutto*.[21] Because foreign students were so numerous in Padua—it is estimated that 6000 foreign students attended the university in the sixteenth century—and so well received, its administrators had little need to associate the annual anatomy demonstration with the rituals of Carnival. Moreover, because Fabrici was developing a natural philosophical tradition of anatomy and wished to incorporate the material stuff of anatomy into longer explanations of natural order and purposeful design, the potential conflation between anatomy and the bodily and bawdy traditions of Carnival would have been entirely counterproductive.

While scholars have pursued the idea that spectators were fascinated by corporeality and human mortality and sought discussions of both in public dissections, the Paduan history of anatomy and its theaters tells a different story. Padua's case indicates the existence of alternative traditions of anatomical inquiry, traditions marked by particular intellectual, social, and pedagogical trends and by specific geographies. Unlike other polities, the Venetian Republic could pay its professors (and give some of them raises), fill its vacant chairs, and protect (somewhat) its foreign students.

## STUDENTS AND TEACHERS

If the anatomical theater was not a place dedicated to literally seeing, it begs the questions of where, when, and how anatomy *was* studied and seen. Such questions are, first of all, pedagogical ones. David Kaiser has suggested promoting pedagogy to a central analytical category in order to refine and expand our understanding of the history of science and its material practices, self-fashioned personae, and moral economies.[22] According to Kaiser, pedagogy "is where the intellectual rubber meets the politico-cultural road," which suggests that pedagogical norms and deviations from them, once characterized, could help to identity the significance of anatomy in the Scientific Revolution.[23]

Although Vesalius is celebrated for his attention to the visual experience of anatomy and the inauguration (in the 1540s) of a new visual approach to studying nature, concern for habits of sensing—and especially with seeing—did not appear as a substantial component of anatomical pedagogy until the late 1580s and early 1590s. Anatomical texts from the late medieval and early modern periods discuss "sensed" anatomy as part of the anatomical method that combines sensual or sensed perceptions with reason, but it was only in the 1580s that a more complicated discussion began to take place about the practical and pedagogical necessity of literally seeing an anatomy. When it did, it was coupled with an explicit and widespread appreciation of the hard work that was required to manipulate nature and with an appreciation of manual, technical expertise. This pedagogical development indicates not that we just need to push back the dates of an already familiar visual regime in the early modern period, but rather that we need a more comprehensive explanation of how a new visual emphasis emerged in the history of anatomy. As student and university records reveal, this new interest in seeing came from institutional changes to the curriculum, pedagogical decisions on the part of professors, and the concerns of students. These might parallel developments in other geographical locations, but Padua's history and its archive afford us the opportunity to submit those developments to thick description (i.e., in the tradition of anthropological inquiry) and characterization.

Though Vesalius praised his own and others' ability to dissect and visibly reveal the structures of the human body, students' records show us that it was not until decades later that this interest grabbed the attention of medical students. In the late 1580s, students ardently began to seek training in the procedures related to surgery and dissection; they focused on ways to isolate anatomical structures and on the need to see them with their own eyes.[24] By the mid-1590s, medical students endorsed the visual orientation of private anatomies. The very thing that Vesalius is credited with—the proof of seeing for oneself—took approximately four decades to become a significant part of the pedagogical culture of the university. In order to explain why, I will look more closely at the relationship between students and teachers.

By focusing on this teacher-student relationship, we begin to understand how the senses—hearing, touching, and seeing—were activated in anatomical education and organized into habits of attention, concentration, and focus. Pedagogy is the art of training and of educating. For anatomists, it includes their goals for courses and their habits in the classroom: their styles of lecturing, demonstrating, and disputing (with colleagues, students, and textual sources) as well as their verbal tics, gestures, and idiosyncrasies.[25] While not all of these elements are accessible from the early documents on anatomy, many of them are. Some anatomy textbooks and, especially, student notes emphasize the pedagogical traits of post-Vesalian anatomists, both those traits that were praiseworthy and those that were admonished. Fabrici's teaching was frequently described as disordered and incomplete, while the lessons of Paolo Galeotto and Giulio Casseri (Iulius Casserius, 1561–1619) were described as comprehensive and clear. Galeotto and Casseri also offered "beautiful" dissections, a description that points to a technical skill that was rarely noted in Fabrici's courses. Galeotto and Casseri were able to maintain an emphasis on the details of anatomical structures, gradually extending their comprehensive demonstrations into areas of natural philosophy. Such particularities, analyzed in detail and over time, provide the foundation for the spectacular appeal of anatomy both inside the walls of the university and beyond.

Students were not the only thoughtful critics of pedagogy. Anatomists were especially mindful of their role as teachers. Following their scholastic predecessors, they assumed that you could not be truly learned or be considered an authority unless you taught.[26] Vesalius chose to thematize the subject of pedagogy in his monumental work, De humani corporis fabrica (1543; hereafter referred to as Fabrica). Though rhetorically sophisticated, his dedication proceeds as a trenchant pedagogical critique against his contemporaries who "scorn working with their hands [manus operam fastidientes]" and who "aloft on their chairs croak away with consummate arrogance like jackdaws about things they have never done themselves but which they commit to memory from the books of others."[27] He placed the "work of the hands [manus operam]" in the context of both current and ancient pedagogical practices: he wished to teach students as Galen did about internal and external human anatomy, urging them "to undertake dissections with their own hands."[28] Paralleling Vesalius, Fabrici was also concerned about pedagogy and the education of his students. It is well known that he published his studies of motion, the senses, digestion, respiration, and generation in relatively short monographs. Less well known perhaps is the reason for choosing such a compact format: it was based in part on students. The brevity, he explained, would keep the cost down and ensure that his students "will be able finally to put them all into one volume and bind them together without any unnecessary loss of text or money."[29] Just as Vesalius introduced a key point—the idea of dissecting with your own hands—through a pedagogical critique, Fabrici's pedagogical concern turned on one as well. Given the inexpensive format of Fabrici's publications, the students could realize his goal of creating a true, complete *theatrum* of "the Whole Animal" (a composite unity made up of the structures associated with each topic). The "paper theater" would serve as a mnemonic device, enabling students to complete their "mental theater" with a clear understanding of the processes of the organic soul as they manifested themselves in human and animal bodies.[30] While both these mental and paper theaters were more complete, idealized extensions of the permanently constructed anatomical theater, it should also be noted that the idea of a theater organized the broader contours of Fabrici's program. In many ways, the various theaters of anatomy—mnemonic, paper, and actual—shaped anatomical education.

In the sixteenth century, the pedagogy of anatomical instruction was determined by several factors, because anatomy was taught to medical students in different ways and in different venues. Although anatomy might seem essential to medical diagnostics and to the investigation of the causes of disease, this use of anatomy was secondary. Anatomy developed in the university setting as a tool for philosophical understanding.[31] This is why Vesalius explicitly located anatomy in the "branch of natural philosophy."[32] In the second half of the sixteenth century, Fabrici worked to emphasize anatomy's relationship to natural philosophy and to situate the study of anatomy in the theoretical branch of the medical curriculum. There, it gained considerable distinction as a field of inquiry.[33] The practical branch of the medical curriculum focused on texts related to diagnosis and the treatment of ailments (most often, Hippocratic works and parts of Avicenna's *Canon*). It included discussions of method and of anatomical, pathological, and therapeutic knowledge, as well as of surgery. Basing their teaching on Avicenna's *Canon*, professors of medical practice, as Paul Grendler explains, discussed the head and brain in the first year; the lungs, heart, and chest in the second year; the liver, stomach, spleen, and intestines in the third; and the urinary and reproductive systems in the final year.[34] The theoretical branch focused on the natural philosophical foundation of the knowledge used in the study of medicine; it tended to consider systematic knowledge, such as the structure of causal explanation used to produce *scientia* and, especially, the purposive or final cause of a condition, phenomenon, or anatomical part.

Natural philosophy was devoted to the study of nature, emphasizing purpose, intention, and design in nature; anatomical knowledge could be characterized as philosophical because it established principles about bodily structures, which were seen as formal expressions of the soul. For example, Fabrici developed his explanation of the uterus by emphasizing Nature and the power of semen:

> To prevent the fecundative power of the semen from being in any way evaporated and to allow it to remain longer in the uterus and be imparted to the whole organ, Nature has confined it and placed it in a cavity, a purse [bursa], so to speak, which is situated near the podex and connected with the uterus. This cavity is furnished with an entrance only, the better to preserve the power of the semen.[35]

Here, Nature had intention, and, following Aristotle's ideas, semen both carried and imparted form. Fabrici used Nature's intention and Aristotle's concept of semen to understand and explain the anatomical structures of a baby chick's reproductive organs: why the cavity connecting to the uterus was there in the first place and why it had only one opening. He then turned his account to the principles of Aristotle's theory of generation and to the role of heat. By the end of the century, Fabrici not only assigned anatomy to the domain of natural philosophy but also ridiculed other anatomists for "moving too far from the foundations of philosophy."[36] As this implies, the pedagogical opportunities that an anatomist had available to him and the subsequent choices he made depended on where he located anatomy with respect to the institutional framework and to the curriculum of practical and theoretical medicine.

## PUBLIC AND PRIVATE ANATOMIES

In addition to featuring Fabrici's research agenda, the venue of anatomical instruction determined the pedagogical style and content of an anatomist's work as well as the expectations of his students. These venues could be public or private. It is difficult to overestimate the importance of the distinction between public anatomy demonstrations and private anatomies, because the public-private split determined so many factors of an anatomical proceeding: the physical location, the audience, the professors (eventually) and the content, the pedagogical goals of the lesson, and the rhetorical key for the presentation. While historians of anatomy have noted this distinction, they have not seen these public and private venues as interactive, mutually constitutive, or products of historical change.[37]

The public anatomy demonstration was an annual event, typically conducted after Christmas and before the onset of Carnival and Lent. These were not only the coldest months of the year (a natural retardant for the processes of decay); they also marked the break between the fall and spring sessions, or semesters. The public demonstration could last up to six weeks, depending on the amount of available material. Before permanent theaters were constructed, this event would take place in a temporary theater, built along the lines of a small stadium surrounding a table. With the construction of permanent anatomical theaters in Padua in 1584–1585 and again in 1594–1595, the importance of the demonstration increased. Not only did nonacademic spectators such as fishmongers and shoemakers come to the second theater, but these annual public events also began to assume civic authority and generate pride in the eyes of administrators and magistrates. The dramatic qualities of the annual demonstration evolved from its intellectual goals and its ritualistic origins as well as from its location in the new theater and its new civic orientation: the theater was decorated with academic insignia, the processional entrances that began the event celebrated the most important members in the audience, and the demonstration itself was sometimes accompanied by music.

In contrast, private dissections (sometimes called "particular anatomies") took place throughout the academic year in local hospitals, smaller classrooms, nearby apothecaries and pharmacies, and the homes of local practitioners and professors. Often closely connected to clinical medicine, they could treat an individual's cause of death (autopsy) and look at specific aspects of anatomy related to ongoing research questions, at medical (rather than natural philosophical) knowledge, and at surgical maneuvers. While public anatomies emphasized philosophical themes and accrued civic significance, private dissections tended to focus on the material aspects of anatomy—students, for example, were especially fond of private lessons when the teacher covered the procedures for dissecting specimens. Private dissections, however, also became a source of suspicion. They provoked fear in the lay community, in part because the procedures for acquiring cadavers remained opaque.

As these details indicate, there were two traditions of anatomy—the public and the private. Never entirely separate, these traditions depended upon one another, sometimes developing shared correspondences and, at other times, using differences to redefine their particular content. In charting the history of the public and private traditions of anatomy, I will return to the subject and practices of teaching and learning, locating their developments regarding the history of anatomy and dissection at the very intersection of teachers, students, and institutions. In their descriptions and reflections, medical students capture the intellectual, social, and emotional dimensions of anatomical inquiry, a range not fully heard in their teachers' publications and their institution's decrees, but a range nevertheless crucial to our understanding of the history of anatomy, its various investments, and its various bodies.

# CHAPTER ONE

# Spectacular Anatomies

## Demonstrations, Lectures, and Lessons

In 1556, when winter had arrived in Padua and the break between semesters was approaching, the students were eager for an anatomy demonstration. Their professor, Gabriele Falloppio, also turned his thoughts
to anatomy.[1] In a letter to the *riformatori dello studio*, a group of two or
three magistrates appointed by the Venetian Senate to oversee the activities and judicial proceedings of the university in Padua and take official
requests to the Senate, he highlighted the eagerness of the students, especially the German and Polish students, for a public anatomy demonstration. He also speculated that if preparations for the annual event were
not soon and evidently underway, these students would leave the university for either Bologna or Ferrara, where such spectacles (*feste*) were regularly held.[2] He then asked the *riformatori* to intervene by writing a letter
to the *podestà* (a city official) in order to expedite the acquisition of cadavers: the body of a criminal who was also foreign (*person ignobile et non
conosciuta*) should be passed "covertly" to the student-assistants.[3]

Beyond the details it offers for the life and career of Falloppio, the letter contains many insights into the realities of anatomical study in the
mid-sixteenth century, after Vesalius's departure from Padua and at the

beginning of what has been called the post-Vesalian era.[4] Especially important is the fact that the study of anatomy was motivated not only by professors such as Falloppio (who wrote letters to obtain corpses through the proper channels), but also by students. It was their eagerness and their perceived threats to abandon the institution that propelled the study of anatomy forward. Falloppio's letter also indicates that both students and professors looked to a range of academic and civic officials to mediate the event's organization, supplying (if not expediting) an adequate number of cadavers for the demonstration. Finally, it describes the event as a *feste*, a ceremony noted for its potentially spectacular appeal.

Falloppio made this spectacle more specific and enticing when, in ending his letter, he mentally arranged the demonstration. He promised "to make a most beautiful anatomy [of the body]" and to combine it with dissections of fish and a monkey; it would last thirty days, not include much reading, and conclude "before the lessons of the new year come."[5] Although the emphasis in such demonstrations usually fell on the reading of a text and the anatomist's ability to engage the audience and produce commentary, Falloppio emphasized the dissection. In this way, Falloppio assured the *riformatori* that the event would last *only* thirty days, ending before the arrival of the new semester. With "hard work and diligence," he promised "to show these hidden mysteries of God."[6] The temporary anatomical theater would display Falloppio and his anatomical knowledge. It would also link that knowledge to his diligence in dissection, that is, to his practiced skill at cutting and showing. Similar to the many demonstrations that had preceded it, this one coupled its didactic purpose—to study human anatomy—with the quasi-aesthetic and religious one of contemplating the "beauty" of the human form. This appreciation of the body was gained and underscored (not destroyed) by the process of dissection. Dissection revealed the secrets of nature (and God), and Falloppio placed himself in the midst of those mysteries.[7]

While the description may sound familiar to us—we would expect an anatomy demonstration to foreground dissection, and many of us would expect the early modern study of the body to have a quasi-aesthetic and religious character—it reveals one instance in which standard pedagogical procedures were modified. Usually the demonstration focused on the reading and explication of a text alongside the body, but Falloppio chose instead to emphasize his skills at dissection and display. The spectacular force of the scene depended not only on the body as a container of divine secrets (*secreta naturae*) but also on Falloppio's ability to reveal that *body* to his spectators by palpating, flaying, cutting into, arranging, and gesturing toward the anatomical parts of the corpse. The manual skills of the surgeon (or even the butcher) were being grafted onto the academic anatomist and cultivated as sources of insight and knowledge.[8] They were derived from a modified pedagogy and mark the mundane but nevertheless crucial origins for the spectacular anatomies of the Renaissance.

In order to understand how Falloppio construed his performance of anatomy—what he thought he was doing and why—this chapter provides a brief history of the public anatomy demonstration, the anatomy lecture, and the private anatomy lesson. These three venues crisscrossed over the course of the sixteenth century and shaped the pedagogical habits of anatomists. In temporary anatomical theaters, the public demonstration made possible the study of the body, but the anatomist, rather than the corpse, dominated the setting. His words, not his hands, were the keys to anatomical knowledge, the true sources of illumination. University statutes decreed that the dissection was performed by the cutter, or dissector (*incisor*), who could be a nonacademic surgeon, barber, or barbersurgeon; a learned surgeon; a university reader in surgery (a position less eminent than the ordinary and extraordinary chairs of anatomy and surgery); or, by the end of the century, an advanced medical student.[9] By the mid-sixteenth century, however, the public anatomy demonstration was described as frustrating for both its audience and the anatomist. At this point, another model for the anatomy demonstration appeared, one that was characterized as an inquiry into anatomy, as an anatomist's revelation of secrets through his own skillful dissection, and as an experience that engaged the body of the spectator—not just his mind or his ears. To understand why these features emerged in the public anatomy demonstration (and why they did not persist), we must turn to Falloppio's demonstrations and his published work, for there, the subtle changes in anatomical pedagogy reflect new interests in anatomical procedures and in structural anatomy, interests that changed the nature of the anatomical spectacle in the post-Vesalian era.

If Falloppio was in the habit of orienting his public demonstrations around his dissecting skill and his skill at showing, he also put these features in dialogue with the importance of practice and experience. Anatomy was best learned from observation, but it was also true that good observation required training. One needed guidance as well as practice with dissection, manual dexterity at isolating (and defining) the objects of anatomy, and experience with dissection in order to manually and visually perceive the particularities of anatomical structure. This was not passive viewing, for it required the *sustained* engagement of the mind and the body. These dimensions and the impressive epistemological and ontological complexity they suggest lay at the heart of Falloppio's modified pedagogy and his spectacular demonstrations.

## THE VENUES OF ANATOMY

In public anatomy demonstrations, anatomists were not usually so assertive about their role in showing—in manually pointing to the parts of the body that needed to be and then were dissected. This was because anatomists were involved in other activities: not *showing* anatomical parts to an audience, but rather *talking* to that audience about anatomy. According to university statutes, the typical public anatomy demonstration was organized by the rector (an elected student position)[10] and two *consiliarii* (presidents of student nations), as well as two senior medical students called *massarii* (a word that derives from mace-bearers or treasurers but refers, in this context, to trained assistants) who were responsible for setting up the event.[11] At student universities, such as those in Bologna, Padua, and Catania, the *consiliarii* were widely recognized because they served on a general assembly that governed the whole of the institution (professors had a separate assembly).[12] The two *massarii* were elected through their affiliation with the student nations. These nations were demographically specific: when a student came to the university, he would matriculate into the nation that represented his geographical place of origin.[13] For the anatomy demonstration, these two student-assistants were responsible for gathering the instruments and procuring the specimens (through official channels).[14] They were also supposed to regulate attendance, not allowing any student to enter who had not matriculated in medicine for at least one year. Once the demonstration began, they were supposed to control the audience. Moreover, criteria for evaluating a dissection were developed around these elected students, an aspect discussed in chapter 5. With these figures playing such active roles, the demonstration embraced its setting. It reflected academic procedures and hierarchies, power struggles within and between student nations, and the consolidation of intellectual and institutional authority.

The anatomy demonstration normally would take place in the winter months and last up to six weeks. According to the 1545 statutes for Padua, the demonstration would proceed as follows:

> The rector and the *consiliarii* are to depute one of the *doctores extraordinarii* who will recite and read the text of the *Anatomy* of Mondino, and one of the *doctores ordinarii*, either of practice or of theory, who will explain [*declaret*] the aforesaid text line by line, and what he has explained according to the text and letter, let him demonstrate [*monstret*] by visual testimony, and verify in the cadaver itself.[15]

According to this description, the *lector* (an extraordinary professor) recited passages from a text; the *demonstrator* or *ostensor* (an ordinary professor from either the practical or the theoretical branch of medicine), who probably translated that material from Latin into the vernacular, pointed out to the dissector, or *incisor*, the parts of the body that should be dissected. The statutes then indicate that a reader in surgery was supposed to do the dissection unless the student-assistants deemed him unacceptable.[16] The statutes offer, somewhat ambivalently, two sources of authority: the *lector*, who channeled the knowledge of the text, and the *demonstrator*, who was supposed to "verify" that knowledge in the cadaver, in other words, make the body illustrate and confirm the text.[17] The statutes tell us that the demonstration was a kind of performance meant to instantiate authority; in this respect, changes in performance will come to illuminate what were changing conceptions (or constructions) of authority for both the *lector* and the *demonstrator*.[18]

By the mid-sixteenth century, some statutory details were no longer respected in practice. For example, the dissection was supposed to follow a lecture and a reading of the *Anothomia* (1316) of Mondino (Mondino dei Luzzi, or Mundinus, d. 1326), a dissecting manual that draws on the anatomical studies of the classical writer Galen as well as on the medieval scholastic tradition of anatomizing.[19] In practice, however, the textual passages could derive from the first *fen* of Avicenna's *Canon* or from Galen's *De usu partium corporis humani*, as well as from Mondino's *Anatomia*. Moreover, as Andrea Carlino has explained for the situation in Rome, the hierarchy shifted so that the ordinary professor, who was more eminent than the extraordinary one, occupied the position of *lector*—a shift indicating that the primary source of authority tended to derive from texts and those who taught from them.[20] Carlino has also found that the positions of *ostensor* and *incisor* were increasingly combined. In Padua, as we will see, Falloppio was in the habit of emphasizing his ability to *show*, indicating another modification to the statutes.

In the early sixteenth century, public anatomy demonstrations were given as introductions to inexperienced students. Jacopo Berengario da Carpi (ca. 1460–1530), for example, who taught anatomy and surgery at the university of Bologna from 1502 to 1527, called the public demonstration a "common anatomy" (*anatomia communi*), not only because it was given before a large audience but also because it served an introductory function.[21] These public demonstrations covered general anatomy and seemed to emphasize both the importance of seeing anatomy—that is, visually apprehending anatomical knowledge—and the problems of seeing: who was literally able to see the structures (a spatial question) was complimented by a sense that a neophyte lacked the training to see (i.e., attend to) *all* the displayed structures. When discussing the mesentery, Berengario da Carpi said that a common anatomy "does not show everything perfectly"; and in discussing the muscles, he insisted that "demonstrating anatomy according to the way it is done in the public gymnasiums [for students], one is not able to demonstrate . . . the great number of muscles nor the bones nor the nerves nor the veins"—for that, one needs "hard work" as well as "a lot of time and a convenient place."[22] Indeed, what one needed, in addition to a skillful teacher, was a private dissection.

Private lessons were a familiar part of a nobleman's education. For example, after 1565, when he was promoted to the first ordinary chair of philosophy, Francesco Piccolomini divided his time between the duties of a professor and the private teaching of noble Venetian *giovani*, many of whom would go on to occupy powerful positions as magistrates and even as doge.[23] His private lessons connected philosophy to the development of virtue, civic virtues, and justice. In the realm of anatomy, private lessons were the occasion for more specialized inquiries and treatments. In such private settings, professors and students could treat the locations and origins of disease and hone one's ability to dissect, manipulate, and scrutinize the body. Although in practice private lessons seemed to have been sporadic, held when corpses became available, they were consistently associated with learning about anatomical particularities, including rarities, abnormalities, and the effects of disease. They also promised to resolve many of the problems of seeing mentioned above.

Private anatomies, as Berengario da Carpi implied, were associated with "hard work"—with the physical labor required to cut into, pull apart, empty, shift, and further manipulate corpses—and with the mental concentration needed to sustain the study of the particulars of the corpse.

By the early sixteenth century, in Bologna (where Berengario da Carpi taught) and in Padua, both public and private anatomies emphasized sensory experiences. In practice, however, they differed. The public demonstration focused on how anatomical knowledge was gained from the senses, what Berengario da Carpi called *anatomia sensibilis*, an anatomy of the potentially sensed parts. This referred to parts that could be sensed even if, during the demonstration, the actual parts were corrupt or not visible in the corpse. This approach was designed to provide an account of anatomy that was based on an understanding of normative anatomy (in place of diseased parts or particular abnormalities, the normative/ healthy ones could be described). It also led to a contradiction, as the demonstration seems to have emphasized the anatomy of the sensed parts and the acquisition of anatomical knowledge through the senses in part because it prevented or frustrated much of that sensory apprehension, especially the ability to touch the parts and see their connections. This was troublesome, a genuine hindrance to knowledge of the body and its parts. While sensory activities, and especially observation, have come to exemplify the innovative developments in Renaissance anatomy and to signify some of the origins of modern scientific practice, it is crucial to understand, and it is generally underappreciated, that they emerged and continued to reside more fully in private exercises rather than in public demonstrations. Though these private anatomies were often unofficial or last-minute events and could take place in classrooms as well as hospitals, apothecary shops, and the homes of professors, they were designated as fruitful sites of experience: with cutting specimens, handling and manipulating their parts, and isolating structures.[24] These anatomies eventually gained recognition and a more definitive place in the curriculum and the institution.

The distinction between public and private anatomies points to two divergent origins for academic anatomical exercises. Private anatomies more closely parallel the tradition of autopsy, which evaluated the individual characteristics of a corpse and typically sought the cause of death (and in this, they played an important role in legal settings).[25] While literary scholars have extended this approach to the academic traditions of anatomy—Jonathan Sawday, for example, suggested that the "autoptic vision" of dissection, once it was applied to the public arenas and theaters of anatomy, encouraged an individual identification between spectator and corpse[26]—this extension usually elides the significant and fairly well-maintained differences between the private exercise and the public demonstration. Instead of treating the anomalous characteristics of a ca-daver's anatomy as autopsies did (and as private anatomies also some-times did), public demonstrations provided a general understanding of human anatomy, as well as a synopsis of the current debates in the field and a few of the connections between anatomy and disease. These points follow the very structure of the medieval anatomy text by Mondino, which was still being used in the sixteenth century.[27] While students may have entertained thoughts about their own bodies and their own state of health, the public venue offered an overview of anatomy as a de-veloping field of knowledge. This public setting also refined their under-standing of academic authority. It reinforced the hierarchy of power and authority, situating the professor and his knowledge at the top and stu-dents in *statu pupillari* beneath him.

In Padua, the public demonstration was often a counterpart to a lec-ture series; they could be combined or held separately, a bifurcated for-mat also used in Bologna.[28] As Andrew Cunningham has explained, these lectures and demonstrations should not be thought of as regressive.[29] They were not intended to be lessons on "discrete items of information" or research exercises that promoted anatomical discoveries. Instead, they had several functions. During a lecture, a medical professor would de-velop aspects of commentary; he might also treat some of the connec-tions between anatomy and disease—which diseases, for example, strike which organs or which complex of structures. In the mid-sixteenth cen-tury, the lecture could be given either by a professor of theoretical or prac-tical medicine or by a professor of anatomy and surgery. At the demon-stration, the anatomist would orchestrate the opening of the cadaver and the discussion of relevant textual passages; that is, he would coordinate the activities of the *lector*, the *demonstrator/ostensor*, and the *incisor*. De-pending on whether he played the role of *lector* or *demonstrator*, he would explain the categories of the body—the three venters, or the spiritual (the head), the animate (the thorax), and the nutritive (the abdomen)— and examine the parts within each. He would discuss issues (in the commentary tradition) related to particular and general features of human anatomy. With his cadavers and his books, the anatomist might attend to the isolation or discovery of new structures, such as Falloppio's uterine tubes or the "doors" (valves) in the veins found by Falloppio's successor, Girolamo Fabrici. Finally, he would discuss structural features of anatomy in relation to function (morphology) and to philosophical purpose. The purposive aspects of Renaissance anatomy have not always been emphasized in current histories of the subject, but the argument-from-design lasted well into the seventeenth century, granting both the subject and its theatrical display a special philosophical charge. If the anatomist also provided the anatomical lectures, we can imagine that the two forums would complement each other, with the demonstrations providing more detail and, if possible, a dissection of the parts discussed in the lecture. If the lecture was given by another professor, then, with the likelihood of rivalry and competition, questions of complementarity are more difficult to assess.

### FORMALITY AND FORMAL CODES

Anatomical lectures and demonstrations took place with notable formality. Their formal codes, or style, indicate a further framework for the spectacular nature of Renaissance anatomies and, in particular, Falloppio's demonstrations.[30] Until the end of the sixteenth century, academics and students populated the event; they were joined by magistrates, usually including one or more of the *riformatori*. Nonacademic people in the audience only begin to appear in the documents from the end of the century and in lengthier descriptions of the anatomy demonstrations held within Padua's two permanent anatomical theaters. For the entire sixteenth century, anatomical events were regulated, formal, almost entirely academic events that included lectures and demonstrations.

For much of the century, *distributio* was the term used to describe the distribution of roles in the demonstration. This distribution of roles reflected and reaffirmed academic hierarchies.[31] In master-student portraits, the distribution mirrors the main sources of authority.[32] Figure 1 features the professor as *lector*, reading from a text while seated in his chair ("ex cathedra") and engaging in a dispute (a formal disputation, or *quodlibet*) with the student, who is placed at the end of the row in front of him. Figure 2 also shows the professor playing the role of *lector* and lecturing from his podium, or chair. Whereas the first image represents the academic

FIGURE 1. Leonardo Legio (Leggi), *Fabrica regiminis sanitatis* (Pavia: Bernardino Geraldi, 1522). Reproduced courtesy of the Houghton Library, Harvard University.

hierarchy by making the professor both larger than the students and closer to the picture plane (and thus more imposing), the second image reflects the same hierarchy by placing the professor above the students. It also enlarges the book in front of him, suggesting that this professor's authority resides in his ability to explicate ancient sources and produce commentary.

These kinds of representational strategies help to clarify the academic hierarchy in less straightforward arrangements than that contained in figure 3. Here, the anatomical scene is organized around the body, stretched out on a table, and beneath the table is a basket, which would contain the leftover parts that would be buried with the cadaver. The academic hierarchy is again clarified through the spatial arrangement of the anatomy demonstration: the *lector*, seated in his chair (centered at the top of the image), reads a passage, while the *demonstrator/ostensor* points to the body and the *incisor* dissects. Though the latter two roles were important, the image emphasizes the overarching authority of the *lector*. In the demonstration-as-performance, his authority was drawn from the books

FIGURE 2. Sebastian Münster, *Cosmographey; das ist beschreibung aller länder*, dccclxiii (Basel, 1598 [1592]). From the Collection of The Public Library of Cincinnati and Hamilton County.

from which he read and from the audience, who watched and listened to him and engaged him in disputations.[33]

The close connection between the anatomist and the text highlights the anatomist's humanist inclinations. In the first half of the sixteenth century, anatomists and professors of medicine (in Padua and elsewhere) were engaged in the restoration of classical knowledge, a development we now call medical humanism.[34] Medical humanism was evident in most aspects of university life. The curriculum, for example, moved a medical student through a series of courses on classical medical texts.[35] Even the understanding of temporary anatomical theaters bore the stamp of

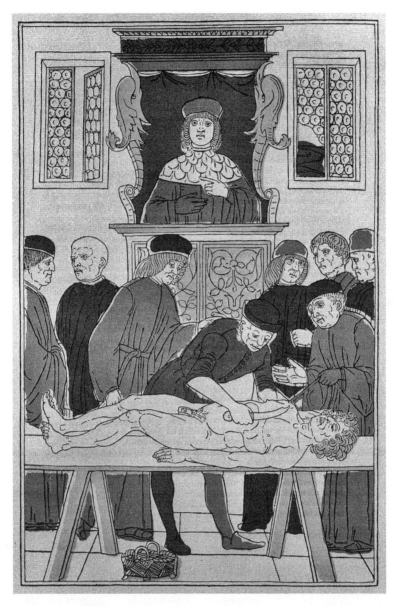

FIGURE 3. Johann de Ketham, *Fasciculus medicinae* (Venice: Gregorio de Gregoriis, 1493). Digital image, M0011499. Reproduced courtesy of the Wellcome Library for the History of Medicine, London.

humanist cultivation. In *Anatomica, sive historia corporis humani* (1502), Alessandro Benedetti (ca. 1450–1512) described a public anatomy demonstration inside a theater: "A temporary dissecting theater must be constructed in an ample, airy place with seats placed in a hollow semicircle such as can be seen at Rome and Verona, of such a size as to accommodate the spectators and to prevent the crowd from disturbing the surgeons with the knives, completing the dissection. They must be able and already have competence from frequent dissections. The seats will be assigned *pro dignitate*."[36] Benedetti provides an idealized portrait of a public anatomy demonstration. The temporary anatomical theater evoked extant classical theaters of Rome and Verona, reflecting Benedetti's Hellenistic inclinations, and served as a mechanism for crowd control. Before permanent anatomical theaters, the public anatomy demonstrations took place in temporary ones, which were built of wood in a circular, elliptical, semicircular, or semielliptical shape, with scaffolding used for those standing and watching the demonstration.[37] These theaters were intended to provide medical students and professors with an educational experience: to learn about anatomy with a lecture that combined textual commentary with dissection and observation. Just as the many anatomical texts that rolled off the presses did, the anatomy demonstration and the anatomical theater participated in the humanist culture of the Renaissance university.

In Benedetti's imagined theater, the source of authority was again the anatomist. It was the anatomist as *lector* who spoke over the *ostensor* and *dissector*, over the students' rumbling, and over the cadaver. While the goal was surely to see the parts of the body, Benedetti's description suggests that it was more important to see and hear the *anatomist*. The process of dissection and the techniques involved in dissecting were not the main focus of the demonstration, for a surgeon—with competence from frequent dissections—was responsible for this work. Moreover, the area with the best view of the cadaver was reserved not for medical students, but for professors and politicians (*pro dignitate*). In its earliest incarnation, then, the spectacle and performance depended on the anatomist himself, for the event was similar to a lecture and was probably familiar to the audience as such (as Ketham's anatomy scene in figure 3 would suggest). Learning anatomical knowledge in this public venue was not a predominantly visual process; instead, it was based on hearing and remembering the framework and details of the commentary tradition, set against the anatomist's own intervention in and organization of that tradition. The spectacle was designed to display and enhance the authority of the anatomist while simultaneously consolidating his power over and through the text.

In the early sixteenth century, the public anatomy demonstration did not intensify the study of anatomical particularities or the anomalies of a cadaver. Pointing to the general (rather than particular) focus of the anatomy demonstration (and to the need to enhance the visibility of dissected parts), Benedetti explained that the criminal and foreign corpses (*ignobiles, ignoti*) used for dissection should be "not thin or obese, [and] of taller stature, so that there may be available for spectators a more abundant and hence visible material for dissection."[38] Cadavers were studied for the examples of general anatomy that they could provide, not for their variations, particularities, or anomalies. Variations were overlooked in the process of obtaining a general, Galenic account of anatomy. Even Vesalius, whose pedagogical enthusiasm is discussed below, called for a body "as normal as possible according to its sex and of medium age, so that you may compare other bodies to it as if to the statue of Policletus."[39] This normative body was male and the object of scrutiny for public anatomies; in private dissections, "which are undertaken very frequently," Vesalius went on to say, "any body can be profitably employed because you will also be able to examine whatever its variations and consider the differences of bodies and the true nature of many diseases."[40]

In the early sixteenth century, human anatomy was studied in the venue of the public anatomy demonstration. There, however, the anatomist used his words (not his hands) to investigate aspects of commentary alongside the decaying corpse. The theatrical nature of the event did not depend initially on the opening of the corpse, on the scrutiny of its parts, or even on some recognized similarity between the corpse and the individual spectator.[41] The public venue, the temporary theater, and the formal codes of the event all directed attention to the professor. For students, it was an introductory lesson, and because students could not scrutinize the parts or even get close to the cadaver, their eyes would turn to the professor and their thoughts, to his words. For the anatomist, the demonstration served professional goals: to obtain knowledge and to establish and enhance his own authority. Benedetti invited prestigious men of the Venetian Republic to one event. In a special preface "to his friends [*ad amicos*]," he describes how he asked them to appreciate those parts of the body most related to their interests. He mentions his friends, Bernardo Bembo and other learned humanists and public officials, who "drove off fright of the dead and left momentarily the cares of the Venetian Republic" in order to contemplate the divine office of the heart and inquire into the "secrets of nature [arcana naturae]."[42]

Because public anatomy demonstrations were pitched to larger audiences (with more eminent spectators) and because the anatomist was the main actor in them, critiques of a demonstration focused on the anatomist's penchant for rhetoric and display. Working as a physician and instructor in the medical colleges of Venice, Niccolò Massa (1489–1569) began his introductory book on anatomy, Liber introductorius anathomiae, by ridiculing the publicly demonstrating anatomist:

> [With God's mercy] I shall write about the remaining parts of the body in another large volume. You should not expect from me [in this little book] a kind of stage of preparation or a seat, where the crowd of spectators may watch you dissecting or demonstrating. Do not expect any such ridiculous thing. I shall not display myself in my writings as an ignorant man [rudem] and not a philosopher.[43]

Massa's disgust is aimed at the anatomist's performance. According to Massa, the public anatomist evoked the figure of the sophist, when he should have emerged instead as a philosopher, a figure of intellectual strength and suitable gravitas. In addition, Massa deployed a common invective against physicians, which associated physicians not only with the excretions of the body but also with the use of rhetoric to deceive their patients.[44] Massa relied on this invective (made famous by Petrarch), insisting that he (Massa) was a philosopher and not a barbarian or a rhetorically deceptive, demonstrating anatomist. Massa extended his critique from the scene of the public demonstration to the printed text, for the rhetorically deceptive anatomist was also the author of grandiose texts, which Massa set in stark contrast to his libretto, or little book.

Massa's invective is noteworthy for another reason, one tied to the kind of introduction to anatomy that he set out to write. His book is a manual for dissection, describing how to proceed from the first incision on through the rest. Drawing on Galen, Mondino, and Avicenna, Massa encouraged the anatomical method of anatomia sensata, which relied on things actually sensed, or sensually apprehended, rather than (as

Berengario da Carpi had indicated) merely potentially perceptible.[45] As such, the right context for understanding Massa was not a public demonstration at the university. Instead, *Liber introductorius anathomiae* would seem to reflect the traditions of learned medicine in Venice, as well as the autopsies that Massa carried out on his syphilitic patients and, later, at the behest of the Venetian *provveditori alla sanità*; it also points to the more private exercises and demonstrations that were routinely carried out, often by Massa himself, in Venice as part of the training and education of the members of the Venetian colleges of physicians and surgeons.[46] In 1536, Massa participated in the annual demonstration in Venice by playing the role of *demonstrator/ostensor*, working alongside a *lector*, a *declarator*, and an *incisor*.[47] Unlike Berengario da Carpi, who used print to provide both an introduction that closely followed Mondino's text and a full commentary on that text, Massa chose to emphasize his interest in demonstrating and showing anatomical findings rather than reading the text of Mondino and responding to topics of commentary.

Massa's work also suggests that "showing" derives from a tradition of anatomical study that is more private than public and more closely associated with practitioners (such as those trained at and teaching in the colleges of physicians and surgeons in Venice). While the anatomy demonstrations in Venice were not designated as public or private, they were intimate affairs. Their audiences were small enough to encourage both an inquiry into causes of death (one goal of an autopsy) and a personal exchange between the participants and the audience. On 15 November 1563, for example, Massa testified in favor of the appointment of Giacomo, a surgeon, to serve the Republic, noting Giacomo's learning in philosophy and in medicine (physic and surgery) as well as his attendance at these anatomical demonstrations: "finding myself with him and discussing the anatomy, I saw and heard that Giacomo has a very good understanding of these matters."[48] The annual anatomy demonstration for the colleges of physicians and surgeons therefore must have been a smaller affair; it was the occasion for a conversation about anatomy that Massa found so memorable, he cited it as evidence in his testimony of Giacomo's competence.

Positioned against the public anatomy demonstration, the private dissection appeared as an intimate and specialized event (Massa's autopsies were done to find the cause of death and to understand the nature of a specific disease).[49] In the 1540s, both the shortage of cadavers and the merits of private dissection were especially clear in Vesalius's teaching and in his publication.[50] In the *Fabrica* (1543), Vesalius explains that

> although a dissection presented privately among a few is undoubtedly to be preferred to a public one, since the supply of cadavers is not sufficient for all purposes, I strive that public dissection be carried on as much as possible by the students so that even those with little instruction—there is always some unskilled person who is willing and eager to undertake dissection at the slightest suggestion—if called upon to dissect a cadaver before a throng of spectators, can conduct the anatomy correctly with their own hands . . . everything to which I have given consideration in this work [the *Fabrica*] I performed several times in the same winter in Padua and Bologna before a large gathering of spectators and in the same arrangement I now present in these seven books.[51]

Vesalius indicated his desire, one that would be echoed by Falloppio, to have students practice the manual techniques associated with dissection and take his book as a guide. He noted that while this practicum was more common in private settings, it should be a part of public anatomies as well. The passage stretches the limits of the public forum, making it more supportive of students by incorporating their practice (or trials) with dissection. We should remember that Vesalius had come to Padua in 1537 and been made a demonstrator in anatomy and a lecturer, positions that were less eminent than either the ordinary or the extraordinary chairs of medicine. From his inferior institutional location, Vesalius reconfigured the anatomist's authority; it is intriguing to think that in the process, he recognized the power that students held at these institutions and sought, therefore, to curry their favor. His authority did not completely depend on lecturing (though the *Fabrica* contains plenty of evidence that this remained important). Vesalius, however partially, began to establish his authority through his ability to show dissecting techniques to students, a "showing" that involved touching specimens, instruments, and, undoubtedly, the hands and bodies of his students.

Vesalius's attention to his students and to questions of pedagogy is a consistent, if sometimes overlooked, part of his anatomical studies. The title page of the *Fabrica* provides an image of a public demonstration that has been retooled and reoriented around the touching and showing of anatomy (figure 4). In it, Vesalius, like Massa, occupies the role of *demonstrator/ostensor*, not *lector* or *incisor*, pointing to the body (and not the

FIGURE 4. Andreas Vesalius, *De humani corporis fabrica* (Basel: Oporinus, 1543), title page. Reproduced courtesy of the Boston Medical Library in the Francis A. Countway Library of Medicine, Harvard University.

text). He touches the womb of the cadaver, which has been retracted in a recent dissection; and he is surrounded by a crowd of people, which is given visual segmentation by the presence of hands. The emphasis on touch is further developed through the contrast between Vesalius as *demonstrator/ostensor* and the *lector* and the *incisor*. The image caricatures the role of *lector*, played by a skeleton, and the role of *incisor*, played by two young men (barbers, barber-surgeons, or surgeons), who sit under the table and whose function is limited to sharpening knives (rather than cutting or handling the corpse).[52]

While this scene has generated a nearly endless stream of differing interpretations, one constant is the assumption that Vesalius has done the dissection himself. Remarkably, however, this is not what the image depicts. While it is surely too speculative to suggest that Vesalius let a student dissect this body, even partially, it may be equally simplistic to attribute to Vesalius the full role of dissector, or *incisor*. Or, if we attribute the role of *incisor* to Vesalius, it must be with the recognition that Vesalius lauded his own ability to dissect and his ability to teach students to emulate him, and the image holds out both possibilities—that Vesalius dissected this body and that Vesalius helped his students to dissect this body (just as his remarks elsewhere suggested). The image presents an argument in favor of doing dissection as an essential component in the process of gaining experience, knowledge, and expertise in the study of anatomy, as many critics have noted. This point, however, is made indirectly (through an implied dissection) and visually (by reflecting Vesalius as a *demonstrator/ostensor* at the moment he is touching an already dissected womb). The image emphasizes "showing," which was one part of the standard pedagogy for anatomical instruction. It was also the part that established the clearest and strongest connection between the anatomist and his students.

Although the image has been interpreted in two opposite ways—as a rejection of book learning or as an imitation of a Galenic practitioner (using knowledge obtained from book learning)—its emphasis on literally "showing" highlights the importance of the broader pedagogical context of anatomical instruction and, especially, private dissections. Vesalius's self-presentation on the title page, however enigmatic, implies that the problem with the public anatomy demonstration as a didactic venue is that it obstructs the student's access to dissection, to the procedures of cutting and manipulating, to the corpse, and to Vesalius himself. The public demonstration impedes the ability to touch and to see what is shown. Not only is it a large venue, but the anatomist, according to Vesalius, is also constrained by the conventions of academic demonstrations and commentary. The venue and practices associated with private anatomies, the image suggests, should be harvested for their unobstructed sight lines, their tactile richness, and their ability to bring together students, professors, and corpses, face to face.

The questions of pedagogy that the image raises were integral to the agenda and argument of the *Fabrica*. Vesalius took a humanist approach to the anatomical tradition, modeling the ideal anatomist on Galen and working his way through Galen's anatomical studies. In *De anatomicis administrandis* (1.3.231), Galen explained that the student "must carefully do everything himself, even to removing the skin"; and again in *De usu partium* (3.98.14-16), Galen famously noted that to learn anatomy, one must "put his trust not in his books but in his own eyes."[53] Equally important, however, was Vesalius' own pedagogical culture, with its traditions and the norms that governed its venues. In the *Fabrica*, Vesalius began by criticizing the format of the public anatomy demonstration and the current pedagogy of anatomists. He ridiculed the anatomist for "croaking [*occinentibus*]" out the lecture from the heights of his chair and drew attention to his "haughty manner of speaking [*linguarum imperitis*]."[54] The croaking reference was to a bird (not a frog), and Vesalius used it to underscore the aspects of display and rhetoric that were evident in the role of *lector*.[55] In the same manner as Massa, Vesalius deployed a rhetorical conceit, sharpening the contrast between the grandiloquent loft of chaired professors and his own expository flatness. Vesalius was unable to praise Charles V because Vesalius's own style was "immature [*ieiuna*]" and "unpracticed [*parum exercitata oratione*]," a rhetorical effect that contrasts Vesalius's discourses with his (refined) manual expertise.[56] Vesalius downplayed his own rhetorical skill, as, paralleling Massa, he had mastered the invective against physicians, turning it against the typical demonstrating anatomist and strengthening its pedagogical roots.

Though the emphasis on touch implies a transference from rhetoric to manual skill—Vesalius describes the latter as refined, polished, and discerning—his interest in resuscitating touch (as a pedagogical topos) appeared again and again, at odds with rhetoric and even language itself. Vesalius indicated that words were only a stopgap, a necessary substitute for when the anatomical part itself could not be seen or touched, and a substitute of a second order, not nearly as useful as illustrations. For example, when he turned to the various forms of bones—those that are rough, smooth, small, large—he provided the following list of analogies:

> It would be difficult, before I have described the bones, to grasp which of them are rough . . . which [bones] recall the shape of the shuttles by means of which coarser threads are woven (such as the bone in the forearm call the radius), or which [bones] we compare to a cube or a tessera (such as the bone in the foot called cuboid because it resembles a cube), or which approximate to the shape of a skiff (such as the bone in the front of the knee joint), or which vaguely recall the outline of the whole of Italy (such as the femur), or which resemble a pin (such as the thinner bone in the lower leg, called the fibula), or which are compared to the beak of the cuckoo (such as the bone beneath the sacrum, known as the coccyx) . . . The very many shapes which bones may have are not likely to be so readily understood by one who has not yet become acquainted with the bones themselves . . . Leaving the distinctions which depend on these matters to their appropriate place, then, let us now consider those which depend on the substance and structure of bone itself [e.g., those that are solid, cavernous, hollow].[57]

This wide-ranging list reflects the variety of skeletal forms and Vesalius's attempt to make them intelligible and memorable. As creative descriptions, the analogies would help the student to remember the bones and their attributes, but, as Vesalius says, such shapes are not likely to be understood until a student "becomes acquainted with the bones themselves." While Vesalius could have emphasized the idea that such a catalog would assist in recognition (not just memorization), he does not.[58] He generates a rich vocabulary and a series of mental images for cataloging bones, but he is careful to mark this knowledge as secondary to the sensual apprehension of various bone structures, surfaces, and shapes.

If we move from anatomical texts to records of anatomical practices, we find that Vesalius's pedagogy was noted for its emphasis on literally showing. This appears to be a product of the overlap between public and private venues of anatomical study—Vesalius had explained that private dissections were more effective sites for learning. Students underscored Vesalius's ability to "show" by contrasting it to the textual (and sometimes tedious) focus of other professors. In 1540, when students in Bologna invited Vesalius to give the anatomy demonstration as a counterpart to the anatomical lectures given by Matteo Corti (Matheaus Curtius, ca. 1475–ca. 1542), then a professor of theoretical medicine, they responded negatively to Corti.[59] According to a Silesian student, Baldassar Heseler, the students disliked Corti's style of presentation: "see how anxious he is to criticize and to boast in unimportant and paltry things, always seeking a vain glory."[60] In contrast, Heseler lauded Vesalius's teaching and, in particular, the way he let students see and touch the dissected and vivisected parts.

Vesalius emphasized the anatomist's touch as an experience directly related to anatomical knowledge, and he then asked students to participate—to experience anatomy themselves in order to gain this knowledge. In the twenty-sixth and final demonstration of the event, Vesalius vivisected a dog in order to demonstrate that when the *nervi reversivi* (the recurrent laryngeal nerves) were severed, the animal would cease to bark. Moreover, he used the dog to elaborate on the relationship between the movement of the heart and the pulse of the arteries. Vesalius said that "you can learn only little from a mere demonstration, if you yourselves have not handled the objects with your hands."[61] As the evening demonstration progressed, Heseler wrote, "I myself saw how the heart of the dog bounded upwards; and when it no longer moved and the dog instantly died."[62] However important seeing the dog was, Vesalius also chose this moment to draw his students into the demonstration, to encourage them to participate by touching the dog, feeling both the movement of its heart and pulse and its warmth. In addition, Vesalius used the moment to stage a potential confrontation with his authority. When the students asked him what he thought about these movements, he replied: "I do not want to give my opinion, you yourselves should feel with your own hands, and trust them."[63] The Italian students ended up mangling the beating heart when they handled it; nonetheless, the entire scene promoted touch not only as a sensory modality but also as an experience related to authority and to the assumption of authority in anatomical matters.[64] This was all made possible by proximity: the triangulated proximity between the anatomist, the students, and the object under study. Although it was not seamless—the Italian students disrupted the experience—it was nevertheless a pedagogical modification, or innovation, in the context of a public anatomy demonstration.

Students readily distinguished these two styles of teaching. The opposition was not solely between the body—the particulars that it revealed—and the anatomical text. Rather, it emerged from the clear contrast between Vesalius's and Corti's teaching. With Vesalius, students characterized effective pedagogy as gestural, tactile, and oppositional with respect to the traditional, sophisticated production of textual commentary.[65] They also characterized his pedagogical style as clear and nearly transparent. As the episode with the vivisected dog indicates, Vesalius tended to play with hierarchies rather than enforce them. Students noted this feature and linked it with the gestural immediacy of Vesalius's demonstration. Vesalius, Heseler said, was "so little communicative."[66] In contrast, Corti presented himself and his material in a more eloquent, refined register, one which produced distance and boundaries between himself and his students.

Students quickly connected Corti's rhetorical skill to the classical tradition of rhetoric. Although rhetoric was an important subject in a Renaissance student's broader humanist education, it is remarkable that it was so easily called forth in the context of anatomical inquiry.[67] In his preface to an edition of Mondino's anatomy, Johann Dryander (1500–1560) declared that he would "admire and appreciate Alessandro Benedetti's anatomy works written in elegant Latin, and Johannes Andernacus's, Andreas Lacuna's, and Vesalius's stylistically well-written relations, if they on their part would admit that Mondino's writings were not without merit."[68] In his notes on Corti's lectures and Vesalius's demonstration, Heseler quoted this passage not as evidence in support of Mondino, but rather as an example of the importance of rhetorical style. Heseler's point was that style was attached to location: elegant Latin was appropriate in monumental books on anatomy. This would be one explanation for the convoluted Latin of Vesalius's *Fabrica* and the gestures and verbal simplicity of his lessons.

Massa had used the same point to criticize the demonstrating anatomist; similarly, Heseler indicated that public demonstrations demanded a different style of presentation, not overly elegant but rather clear, simple, and transparent. Public demonstrations should be delivered in the *stile chiaro*, the clear style. Corti had begun the first lecture of the series with a discussion of the causes of anatomy, following the standard format of the *accessus*.[69] Corti then argued that Galen offered several causes, while Mondino, in his own reading of Galen, had found only three. Corti stated: "as regards what Mondino says, that he does not want to use a high style [*alto stilo*] in these things, I certainly do not know what he means by style. To be sure Mondino has not kept the order of Galen. For style and method in anatomy is to follow Aristotle's and Galen's method and order."[70] In a marginal note, Heseler declared that Corti had misunderstood the meaning of style. Style had nothing to do with Aristotle's, Galen's, or Mondino's methods. Rather, style signified "a manner of speaking":

> The opinion and intention of Mondino was that he did not want to use embellished and refined phrases but he wished to explain the anatomy simply in ordinary language . . . the essential thing is to teach the contents and to speak more clearly than elegantly. We cure with things or herbs, not with verbs.[71]

Elegance—the potential effect of many figures of speech—could and should be sacrificed for the sake of clarity in classroom discussions of anatomy. Though Heseler's pedagogical preferences and his attention to rhetoric may have been more developed than those of his peers, his remarks link anatomy and rhetoric, tying clear style to the content of the demonstration and the public context of its delivery.

Both Corti's comments and Heseler's critique refer to an opaque passage in Mondino's text. In his introduction, Mondino explained: "I therefore purpose to give, among other topics, some of that knowledge of the human body and of the parts thereof which come of anatomy. In doing this I shall not look to the high style [*stilum altum*] but shall merely seek to convey such knowledge as the chirurgical usage of the subject demands."[72] In opposition to the high style, a clear style was used to convey information in straightforward language, without seeming to rely on rhetoric or even language itself. In the mid-sixteenth century, anatomists turned to the idea of a transparent language, associating it with both sight and touch. Jacques Dubois (Jacobus Sylvius, 1478–1555), the Parisian scholar and professor of medicine, explained in his publication on Hippocrates and Galen (1555) that seeing and touching were better than hearing and lecturing for gaining an understanding of medicine; he also declared that words were artificial and that he had hopes of getting at the reality of things "sine nominibus," or without words.[73] For Mondino, the presentation of surgical knowledge demanded simpler locutions as well as gestures that clarified the proper way to cut bodies and set broken bones. It was no less rhetorical but, as a style, it connected loosely to the manual, gestural, and practical features of anatomy and surgery lessons.[74] Although Mondino's description seems to limit anatomical knowledge to knowledge that is useful to the surgeon, the clear style began to receive more attention in descriptions of public anatomy demonstrations, as Heseler's remarks on "ordinary language [*familiari sermone*]" indicate.

In the post-Vesalian period, the refinement of the tradition of public anatomies was in part a refinement in style: the formalities of the public event were eventually combined with a more eloquent style of presentation. By midcentury, ideas about style began to splinter. The private venues of anatomy supported specialized study, more exchange between the anatomist and the smaller audience in these venues, and tactile and visual experiences, and these features were part of what motivated students to endorse a clear style. Public anatomies served introductory functions, and because they were recognized throughout the academic community (rather than merely among a select group of medical students), they adhered more closely to the formalities described in the university statutes, even when both teacher and student were frustrated with those formalities.

The public and private venues of anatomy, however, began to overlap. This can be detected in the new interest in touch, typically a feature of the private setting, and in the clear style, usually a feature of anatomical presentations that treated aspects of surgery. These attributes were sometimes a part of public anatomy demonstrations, where they emerged in modified pedagogies. For example, Realdo Colombo (1516–1559), who was Vesalius's successor in Padua, referred to Vesalius's pedagogical and performative tendencies.[75] In *De re anatomica* (1559), Colombo maintained that the extraordinary nature of the laryngeal complex had not been fully understood during Vesalius's demonstrations in Padua: "Vesalius was accustomed always in public to dissect and to show" the laryngeal structures in cows, and they will be made clear "if you will inspect [*inspexeris*] the human one, [which] preserved in a consistent state all of the small bones (that are doubted from a distance)."[76] Colombo indicated that Vesalius used the showing in his public demonstration; Colombo also noted the confusion that resulted from the physical distance imposed by the public setting of anatomical instruction. Clarity and recognition could only be derived from inspection, the up-close examination of the preserved laryngeal complex that required both touch and sight, an aspect that Colombo pursued through his dual interest in vivisection (the dissection of a living animal) and in private venues.

The public arenas of anatomical study still elicited Colombo's interest, but they also presented difficulties. Their spatial arrangements, their pedagogical rituals, and their formality were experienced as constraints. Colombo dedicated a significant part of his research to vivisection, a procedure that required proximity, the careful touch of the anatomist (using manipulations practiced and perfected through trial and error), and visual scrutiny by his audience. Vivisection highlighted the merits of the private lesson, suggesting that alongside classical precedents and Colombo's noted interest in the Alexandrian anatomists Herophilus and Erasistratus, pedagogical conditions influenced his research agenda and his anatomical practices. In his vivisection of the thoracic muscles in a dog, Colombo wished to show how the exterior muscles draw upwards as the interior ones moved downwards but, as he explained, "this motion I frequently observed in living dogs, dissected; by dissections, that is to say, which were made in house [and] not in public; there [in public] all things can not be considered [as] closely by listeners and spectators as [they can] in private dissections."[77] Vivisection, which uses a living animal, requires proximity to the animal because the movements are often subtle and always fleeting, and, by the end of the century, vivisection had become an occasion for displays of technical expertise. Heseler's remarks about Vesalius's vivisection reflect the growing interest in touch and in what kinds of shifts are possible when the venues and the pedagogies of anatomical study begin to overlap.

Through such changes, public anatomy demonstrations took on a new spectacular appeal, distinct from the lecture-oriented events that were the mainstay of academic culture. By the 1550s, the public demonstration was not solely or always used to offer a general introduction to anatomy. Instead, it began to present aspects of current research, and, as a result, the public demonstration became an increasingly fraught venue. Impediments to success were related to proximity, sight lines, and noise levels, for the public venue, as Colombo noted, could draw crowds of three hundred people or more.[78] Actually showing anatomy and encouraging its visual and tactile appreciation, features consistently practiced and refined in private dissection, now also became topics in public demonstrations—a way of thinking and talking about the spectacular potential of an anatomy demonstration as distinct from a lecture. As a result, the public demonstration became less stylistically formal and less hierarchical. Students were encouraged to approach the anatomist and the specimen and to touch, handle, and scrutinize the parts. These moments of improvisation might degenerate into physical comedy, as Heseler's account of the antics of the Italian students revealed, but they also constitute important, though less explicit, episodes in the training of anatomical inquiry. They depend on proximity, touch, and sight, and they become increasingly frequent in the records of medical students.

### GABRIELE FALLOPPIO

As did Vesalius, Falloppio often modified pedagogical norms so that he could "show" anatomy, publicly performing the role of *demonstrator/ostensor*. Falloppio came to Padua in 1551, when he was appointed to a chair in "surgery, simples, and anatomy" and was responsible for providing public anatomy demonstrations.[79] In 1554, as the annual demonstration was being organized, it was decided that although Falloppio held the chair, Vettor Trincavella (1496–1568) would provide the ordinary lecture for practical medicine, that is, reading for the demonstration, and Falloppio would provide only the cutting and showing (*a tagliar et mostrar solamente*).[80] While the letter outlining the structure suggests that the roles of *lector, demonstrator/ostensor*, and *incisor* could be distributed among three people, two, or only one person, the practices at this demonstration reveal further detours. On 25 January 1555, the actual demonstration took place with Apellato reading the text of Mondino while Trincavella explained and showed it (*esporlo e mostrare*) and Falloppio dissected the body.[81] When it came to the demonstration, the "order was observed" except that Apellato did not do his part very well:

> The first [reading] was quiet, the second early on interrupted. Coming to the reading in the Theater and with great difficulty, he was heard the first time, and coming to the showing, the students were so dissatisfied [that they] did not remain; the second [lesson] was totally interrupted, nor could anyone speak a word because of the noise of the students who were shouting, "we want Falloppio" . . . And so the lesson went no further.[82]

The students objected initially to the reading and then to the reading and showing of anatomy. Their objections helped to establish Falloppio as a performer and to suggest that his dissection (and explication) would

trump both Apellato's unsuccessful reading and Trincavella's explanation and showing. In this episode, the spectacle's appeal is embodied in its
joint emphasis on the anatomist and his dissection of the corpse. Rather
than the professor with his book or his voice, the anatomist and his proximity to the corpse augmented the anatomist's power and control over the
scene.

The episode reveals that the anatomy demonstration continued to be
organized around the three roles of *lector*, *demonstrator/ostensor*, and *incisor*; these positions were not always collapsed into one role and played by
the anatomist (as Vesalius had wished). Consequently, Falloppio's interest in the techniques of dissection, in cutting and manipulating the
specimen, should be seen as the result of the demonstration's format:
Falloppio was supposed to focus on dissecting and not on reading or on
commenting on the text, actions which were done by Apellato and Trincavella, respectively. In addition, this episode demonstrates how powerful the students were in shaping the history of anatomy. Students wanted
Falloppio as their teacher; they also wanted a teacher who focused on
dissection. The administration granted their wishes, motivated in large
part by the need to keep foreign and domestic students enrolled at the
university.[83] A similar thing had happened before, and the situation was
only resolved when Falloppio took over. In this case, administrators asked
that it be conceded that "we make the anatomy according to the way [it
was done in] the years past," by which they meant that "the Excellent
Falloppio" would resume his former roles of reading and demonstrating
("cutting and showing") anatomy.[84] The administration's willingness to
listen to the students and respond to their complaints reflects the eco-
nomic constraints of the institution and the growing importance of anatomical study to the reputation of the institution. As their letter noted,
the university was "the most famous in all the world," so much so that
"many students remain here until the end of their studies and many also
come from other universities only for this [anatomy]."[85]

Falloppio's attention to literally showing anatomy during a public
demonstration correlates with his interest in basic morphology, that is,
the relationship between anatomical structure and function (which
Galen called "action"). This interest was fueled by Galen's *De usu partium* (and its compression in *De juvamentis membrorum*), which were
available by the early sixteenth century (in editions from Pavia and Venice).[86] In *De usu partium*, Galen provided an account of the structure and

function of the parts of the body, relating them to natural philosophy and metaphysics. He proposed three categories for anatomical inquiry: structure, action, and use. "Structure" refers to the material organization of individual parts, and "action" to the function of those parts. The two were closely tied, for both ancient and Renaissance anatomists considered function to follow from the form, or structure, of an individual part. The definition of "use" is more complicated. In the last chapter (*epode*) of *De usu partium*, Galen explained that use or usefulness refers to the suitability of a part for its action, that is, the special characteristics of its structure that enable it to function as it does.[87] The category of use pointed to the skillfulness of Nature and to the elements of design. It drew attention to the principles of the organic soul—those vital functions such as motion, digestion, respiration, and reproduction that were tied to the body and its organic composition—and to the philosophical importance of anatomy.[88]

By covering these three categories, Renaissance anatomy was deemed useful to physicians and especially to philosophers. Some commentators on Galen and Mondino, such as Matteo Corti, said that anatomy was also useful to theologians because it studied the human body, made in the image of God.[89] This was also voiced by popular medical writers. The prolific empiric Leonardo Fioravanti (1517–1588), for example, made this point the primary reason for his extended discussion of anatomy in *La Chirugia* (1581). More routinely, however, anatomy primarily figured in the studies of physicians and philosophers. Basic features of anatomy might help physicians understand the location of diseases and their accompanying symptoms (and thus help with treatments). Physicians needed anatomical knowledge about the human body in states of health and disease, using both conditions for clinical purposes, to derive norms and deviations from them. Anatomy was often considered indispensable for surgeons (and for physicians, although it was difficult to combine with humoral theory).[90] For philosophers, anatomy was thought to provide answers to questions the philosophers posed about nature. Galen insisted that the "one very great advantage which we gain from this work [*On the Usefulness of the Parts*], not as physicians, but, what is better, as men needing to understand something of the power responsible for usefulness, [is] a power which some philosophers say does not exist at all, let alone its providing for animals."[91] For philosophers, anatomical investigation went beyond merely that research thought to be essential to questions about disease and treatments. Part of the time, Galen studied anatomy from practical contexts, such as the wounds of gladiators, but he contrasted this observation with the kind of sustained study required for anatomical understanding. As Andrea Carlino has pointed out, the "occasional" anatomy that Empiricists treated during the exploration of wounds was, for Galen, "utterly futile."[92]

For Renaissance anatomists, the philosophical dimensions of anatomy— its purposive aspects—became increasingly important, and the presentation of these dimensions appeared to demand a higher rhetorical register. This requirement was already evident in Galen's works. The final chapter of his work was called an *epode*, Galen explained, because it resembled the epode that melic or lyric poets used "to chant standing before the altars . . . singing hymns of praise to the gods."[93] The category of use, even for Galen, demanded a more eloquent presentation. For philosophers, the presentation of anatomy included a commitment to a wide-ranging project of inquiry (not merely the anatomical basis for a disease), an appreciation of the body as a reflection of God, and, often, the pairing of philosophy and poetry.

According to this Galenic program, Falloppio's emphasis on cutting and showing is striking because it dilates the account of structure.[94] In a typical demonstration, the material, or structural, features of anatomy were treated at the beginning, a phase called *historia* (or *observationes*).[95] *Historia* refers to *sensata cognito*, a knowledge based on sense perception or on (direct and indirect) observation, and in a public demonstration it was followed by a treatment of actions and uses.[96] Falloppio emphasized and elongated the first segment, refining and extending the *historia* of anatomy. The same procedure is evident in his major publication on anatomy, *Observationes anatomicae*, where he claimed to provide an account of that first phase of a demonstration. There, he promised to treat the following: first, what he *observed* of the bones; second, what he *watched* of the muscles; third, that which he *saw* of the veins, arteries, and nerves; fourth, that which *is to be observed* in the abdominal and thoracic cavities; and lastly, what he noted concerning the parts of the head.[97] As Falloppio explained, he wrote this book around 1557 (four years before its publication) at the insistence of Pietro Manna, a humanist physician (*medico*) from Cremona who served as the personal physician to the Duke of Milan before Manna's death (ca. 1557–1561). Falloppio noted that Manna had named Berengario da Carpi as the first "restorer" of the anatomical art and Vesalius as the anatomist who had "perfected" it.[98] Falloppio was a humanist, following the main tenets of Galenic medicine and the anatomical tradition that stretched from Galen to Berengario da Carpi and Vesalius. Equally, Falloppio was influenced by his contemporary pedagogical culture, for it was in that culture that he developed his own habits and a pedagogical style that promoted "showing" anatomy. Included in that showing was a demonstration of manual dexterity and a call to observe structure.

This emphasis is instructive, because it helps to establish the kind of spectacle that was possible for anatomy demonstrations in the mid-sixteenth century. Though strongly Galenic, Falloppio's publication did not elaborate on the philosophical utility of anatomy (or anatomical inquiry) by exploring the Galenic category of use. The *Observationes anatomicae* was limited to "the processes and appendices that I [Falloppio] had observed"; Falloppio proposed a future work on the topic of *anatomical speculation*.[99] The distinction between observations and speculations implied a distinction between *historia* and descriptions, on the one hand, and natural philosophical dimensions, on the other. The latter were called speculations not because they were probable, but rather because their character was more universal, that is, a universal account of anatomy in relation to soul. Falloppio implies that this has never been done before, suggesting perhaps that he made an early attempt at a more sustained treatment of (Galenic) anatomy, especially the philosophical ideas in *De usu partium*, in relation to Aristotle and natural philosophy (a project later advanced by Falloppio's successor, Girolamo Fabrici).

Falloppio noted the "marvelous" organization and design of nature as it appeared in the dissected corpse. However, he did not write enthusiastically or at length about the natural philosophical, teleological, or other abstract dimensions of anatomy. For example, during his public demonstration of the muscles of the lower abdomen, several men in the audience raised objections.[100] The nature of their demurrals is telling. The first objection was that the parts did not have the character of muscles, so they were not distinct muscles; the second was that the right number of muscles were not available—they couldn't find the fifth pair of muscles "created by nature"; the third was that the parts had no clear function. In his book, Falloppio provided more details about this third complaint, such as mentioning the spectators' inability to work out how these muscles functioned to raise a part of the body or to compress the surrounding vessels.

The *structure* of these muscles was important because it helped the anatomist and his audience to understand their *function*.

Quoting a line from an unnamed poet, "Profitable is the competition between mortals," Falloppio then proceeded to respond to each of these three objections. The first complaint was that Falloppio did not understand the definition of muscles, to which he replied that this was a kind of "smokescreen [*tragelaphum* or, in Latin, *hircocervus*]":

> In truth they know it [the definition], also I know it and before me, Galen in the first book on the motions of muscles, said "the muscle is the immediate organ of voluntary motion." I say that this definition adapts itself very well to these small muscles and that they are truly organs of voluntary motion and that they spring obliquely in the lower part of the inferior abdomen for part of a voluntary compression.[101]

In the context of a public demonstration, Falloppio thus engaged in a formal dispute with his colleagues. He presented not only the abdominal muscles and their functions but also his knowledge of the natural philosophical tradition that derived from Galen, and he framed the study of muscles as a study of motive soul. Noting that this issue was a mere "smokescreen," however, Falloppio suggested that the definition and its implied connection to natural philosophy be tabled (or perhaps better posed in his future work on anatomical speculation). He cast the first part of this disputation as merely a (statutory) requirement for public anatomy demonstrations, rather than as an integral part of his current inquiry. When responding to the spectators' other objections, he referred to his experiences in dissecting and studying many cadavers and then to his study of Galen's work on the dissection of muscles.[102]

Though they were not directed at natural philosophy, Falloppio's anatomical studies were productive. Not only did he continue to catalog the parts of the body and to treat function, but, as a result of his interest in structure, he also developed new techniques for isolating structures. During the annual demonstration of 1553–1554, Mattia Guttrich, a physician of German descent who was practicing in Venice, brought the head of a seal or dolphin or large fish (*foco*) for Falloppio to dissect. Praising Guttrich for being learned in the discipline of philosophy and medicine and for knowing Latin as well as Greek and Hebrew— "he was able to penetrate the heart of medical cures with the force of

fire"—Falloppio then recorded his dissection of the eyes of this fish on
29 December 1553:

> He, Ruberto Phinch [a student-assistant], who I love as a son and who has
> become learned in philosophy and medicine and the art of medicine,
> brought it [the head of the large fish] to me . . . Receiving this head, I then
> began to cut it and dissecting it, I observed that this animal moves the
> eyelids and opens the eye . . . [and] the agents of movement are four red
> muscles . . . I showed publicly in the anatomical theater [these] structures
> and then instructed by this observation, I began immediately to research,
> in an ox's eye, a similar formation . . . made more learned by this experi-
> ence, I found also in the human eye a small and subtle muscle [that simi-
> larly originated and controlled movement].[103]

This public anatomy demonstration was thus the site for research and
for the presentation of results. The fish dissection prompted Falloppio to
go on to study the eye; he began this research with a vivisection and then
developed a technique for isolating the muscles and other structures of
the eye. These structures are difficult to separate out because they are
stable only as long as the fluid in the eye remains intact; if the mem-
branes are punctured, the fluid leaks and the structures collapse. Having first analyzed the muscular structures of the fish eye, Falloppio then
wanted to research these structures further; he sought a similar formation in an ox's eye before he turned to the dissection of a human eye. In
addition to discussing the development of such a careful technique, proceeding from animal to human parts, Falloppio described how he stabilized the structures of the eye in order to scrutinize their composition. The
method, moreover, is collaborative, derived jointly from the interaction of
an anatomist, a local practitioner, and a student. Though complex, the
passage suggests how Falloppio's demonstrations staged his presentation of structures and his techniques for isolating them; it also provides
an example of a process by which students learned to see.

Falloppio's pedagogy emphasized structure: the ability to isolate it
and to perceive it. Nevertheless, he dramatically rendered not only those
moments in which he succeeded, but also those in which he failed.
Scholars have seen signs of Falloppio's humility in his stories of failure,
but these accounts also emphasize the importance of training and the
accumulation of experience. There were two basic kinds of failure in the

*Observationes anatomicae.* The first was the failure of other anatomists to observe keenly. Falloppio said that his personal experience with dissection was one reason why he decided to publish this work:

> There are many anatomists who writing and professing in public or in private, attribute to themselves the finding of many things that by other anatomists have never been seen; ornamented with their pens like Aesop's horn, they expound and go on divulging [these findings] from observations that they have poorly taken; [and] listening to the lessons of others or of myself or of my disciples, they attribute these things to themselves, showing them at times in a distorted and erroneous way.[104]

Just as previous anatomists highlighted the merits of their work by drawing attention to the rhetorical spin cast by their colleagues and predecessors, Falloppio also addressed the errors of his colleagues by way of rhetoric, noting that their errors were covered over by eloquence and rhetorical ornament. Moreover, Falloppio highlighted two additional features. Not only did he flag the academic pastime of attributing originality to claims previously made by others, but he also introduced inadequate observation as a potential difficulty. Some of the errors of the anatomists derived from observations that were "poorly taken." These distortions and errors could be identified at the point when the anatomists attempted to actually show their anatomical findings. In contrast, as the above passage implies, good observations lack distortion and require practice, that is, experiences accumulated over time.

The second kind of failure that Falloppio addressed was his own. In his treatment of the muscles, Falloppio began with a debate on whether muscles were singular or plural in nature.[105] While he opts to treat this at greater length when he has more time, the question lingers, emerging again in his treatment of the dorsal muscles. Falloppio's description of the dorsal muscles reads as though it were a confrontation between the anatomist and a resistant body. Explaining that there was no agreement between anatomists on the structure of these muscles, he states: "Confessing my thoughts frankly, it seems to me that in treating this disordered matter and the confusing chaos of muscles, I desire a guide so that I could follow the dissection clearly made before my eyes, reducing the parts to a number and a certain order."[106] Other texts and other anatomists had failed to serve as this guide. Falloppio notes that another anatomist might take exception to his request for a guide, but for Falloppio, a guide remains necessary "because of the infinite proximal and distal [insertions] that are contained in this chaos." "In reality," he concludes, "it is like a labyrinth of many streets; above all, I say to you with frankness that in this, I return to observation."[107] This passage describes Falloppio's encounter with a recalcitrant body, for what he saw when he looked at the dorsal muscles was chaos, a labyrinth of many streets (not unlike Venice) and an infinite number of fibers. He wishes he had a guide (text or teacher) to follow, either a map to help him navigate or someone to lead him through the labyrinth. In this setting, Falloppio conceives of failure as an inability to distinguish the muscles manually and visually. Falloppio can access the conceptual categories of number (singularity, plurality) and of kind, but these don't make it any easier to distinguish and trace complicated structures. The passage captures Falloppio's frustration, but it also dramatizes the anatomist in his nascent study of structure, when he attempts to isolate structure, the basic building block of any anatomical account. Indeed, the passage is framed by rhetorical gestures of sincerity: he "confesses his thoughts frankly" and speaks to his readers "in all frankness" about his honest intentions, but within the "confusing chaos" of the dorsal muscles, where does one structure begin and another end? Failing to resolve the crisis, Falloppio recommended a return to observation.

This episode insists on training. It tells the neophyte what he will see at the beginning of his training (namely, primordial chaos) and at the end of his training (the intricate paths charted by the proximal and distal fibers of the dorsal muscles). It is also instructive because it reveals Falloppio's rhetorical tendencies. Presenting the specimen as a body that resists clarification (i.e., a body for which there is no sufficient map) highlights Falloppio's manual skill: dexterity and strength are required to wrestle with the refractory body. The anatomist then emerges as one capable of revelation. It was not the inexplicable, awesome revelation of the kind generated by sanctified bodies and miraculous signs. Rather, revelation was the product of hard work. As Falloppio wrote in his letter (which began this chapter): "I will use hard work and diligence to show these hidden mysteries of God."[108]

There were different means to manage the primordial chaos and other ways to spin the presentation of anatomy. The muscles gave Girolamo

Fabrici, Falloppio's successor, trouble as well, and in his *De motu locali animalium* (1618), Fabrici turned to the philosophical dimensions of anatomy:

> You may wonder, reader, that I do not describe the muscles, as did Vesalius in all of his works, and Galen in the book, *On anatomical procedures* . . . whereas Vesalius and Galen place things before the eyes of spectators . . . [I wish] to teach the goal for each, which are the muscles, their actions and uses . . . For if one inquires with simple dissection, and in this way enumerates the sequences first, second, and third, it results in more confusion than if one notes the utility of the muscles. And when we inquire into causes and into uses . . . we commit the exact number of muscles to memory.[109]

Falloppio worried about distinguishing one structure from another, while Fabrici registered confusion about the number of anatomical structures. For both anatomists, the hard work of dissection was not enough to bring clarity to the structure or function of the muscles. Dissection was aimed at isolating the structures of the muscles, but Fabrici diminished its import by explaining that Vesalius's work in the area was both simplistic and confusing. Fabrici consistently moved from the simple to the complex, from the material aspects of anatomy to its natural philosophical significance. In order to do so, he often narrowed the number of structures in his account of *historia*.[110] While there are several explanations for this, which will be discussed in the upcoming chapters, the reason stated here is pedagogical. Because the plethora of structural components had become confusing, Fabrici redirected his students from structures per se to the categories of anatomical knowledge, that is, to systems of organization in which structures were known and remembered according to the uses of the parts. The categories of singularity and plurality, which informed Falloppio's treatment, shifted (for Fabrici) to those of use, the most abstract and philosophical of Galen's categories for the study of anatomy.

The relationship linking Falloppio to Fabrici has not been the subject of historical analysis, in part because Fabrici's anatomical studies seem so different from those of Falloppio. Fabrici, in his anatomical studies, treated aspects of purpose and design and sought to transform anatomical inquiry into a natural philosophical enterprise that was more strictly and coherently Aristotelian than ever before.[111] As Gianna Pomata explains, Fabrici's studies rely on a notion of *historia* drawn from Aristotle: for Fabrici, *historia* did not include all of the particulars that could be summoned on a topic, but rather those that were relevant to the final cause.[112] Because of this, Fabrici's work elaborates on ideas of divine purpose and abstract causes, narrowing the appreciation of structural characteristics yet extending their significance. To explain Fabrici's interest in and turn toward Aristotle, scholars have cited the local context of Padua, which buzzed with Aristotelian studies and debates on the soul at precisely the moment when Fabrici entered it, and then ascribed his transition to a version of one of the following departures: from Galen to Aristotle, from structural anatomy to natural philosophical anatomy, and from Vesalian to post-Vesalian inquiry.[113] Pedagogical traditions, however, tend to insist on continuity. To both Falloppio and Fabrici, the pedagogical context revealed the limitations of a study of anatomy that focused so fully on structural composition (or so fully on Galen's first two categories of anatomical inquiry). Falloppio hesitated over how the dorsal muscles were structured, wondering where one muscle ended and another began. He described more effective techniques to isolate parts; he emphasized training; and he celebrated the accumulation of experience. Addressing similar problems, Fabrici recommended memorizing not the many structures of the human body, but instead the principles that explained anatomical formations across a wide range of species. Fabrici offered his students a novel mnemonic, a way of remembering anatomical particulars according to the similarities and differences in the anatomies of human and animal bodies. He underscored the relationship between structure and function, but he organized his inquiry according to the category of use, the "first principles" that explain the anatomy of every animal.[114] That organization indicated the emergence of a different kind of spectacle, a tradition of spectacular anatomies that moved up toward philosophical speculation and contemplation rather than down toward the interwoven sinews of the corpse.

## CONCLUSION

The origins of spectacular anatomies derive, on the one hand, from the university statutes, procedures, and codes that organized the annual public anatomy demonstration and, on the other hand, from the modification and adaptation of those normative dictates. The university statutes cover many of the details of this annual public event. The cadavers were to be the bodies of criminals, ignoble and also foreign (i.e., not from Padua or Venice). The demonstration required the work of the student rector, two of the presidents of the student nations (*consiliarii*), and two senior medical students (*massarii*). Managed by these figures, the event was to proceed in an organized and controlled manner, moving through the acquisition of corpses, the demonstration itself, and then the burial of the remains (which followed the completion of the entire course and for which a fee was collected from the spectators). In practice, the event took up a range of topics and developed different focal points—such as Colombo's vivisection of the laryngeal (or voice) complex, and Falloppio's account of the muscles of the eye—and it seems to have emphasized the position of *lector*, given that ordinary chairs (rather than extraordinary ones) occupied it. Between protocols and practice, these demonstrations were understood as occasions (public within the academic community) that stimulated the eyes, ears, and mind of the spectator, and as events that staged the diligent toil and struggle of the anatomist in attempting to discover and reveal secrets about human anatomy through dissection.

Falloppio's teaching in Padua emphasized the observation of structure and the intense focus needed to undertake the manual work of dissection. Falloppio relied on training, the guided experience that would eventually translate into expertise. This training took place in private exercises and in public anatomy demonstrations. In public demonstrations, the first phase, *historia*, was probably longer than it had been in the past, because it treated at length the handling and scrutiny of anatomical structure as well as its verbal description. Falloppio seems to have had a special talent for this kind of "showing." The formalities of the public anatomy demonstration had to shift in order to allow students and professors to work more closely with each other and with their specimens, as the example with the large fish and Falloppio's friend, Mattia Guttrich, illustrates.

While a focus on anatomical structure seems both basic and essential, it introduced certain problems into the study of anatomy. As Falloppio studied the details of anatomical structure, he confronted a dizzying array of parts and troublesome irregularities. How was a professor to teach all the parts of the human body to students?[115] When was it useful to treat anomalous and abnormal structures in addition to normative ones? Was the private dissection or the public demonstration better suited to these tasks? These questions were further complicated by noisy students as well as by a lack of cadavers and by heat, which increased the rate of decay, putrefaction, and stench.[116] Such problems helped to define the objects of study, the procedures and method for acquiring knowledge, and the aesthetic and rhetorical keys for presenting that knowledge. Therefore, these problems shaped the production of anatomical knowledge.

When Fabrici began to teach anatomy, he struggled with *historia*, the phase in which an anatomist was meant to literally show and explain the structures of the topic under question. Fabrici's comments about *historia* and his decision to focus his research on processes (thereby limiting the number of structures that needed to be treated in any given demonstration) suggest that for both public and private anatomy lessons, there were too many structures, an abundance that made the first phase of a demonstration confusing to students. Fabrici thus looked instead to the category of use, in other words, to the philosophical importance of anatomy. In Fabrici's hands, anatomy would become a specialized inquiry, a natural philosophical enterprise, and a topic to be treated publicly in a more eloquent style. Influenced by the specific venues of anatomical study, these pedagogical and rhetorical proclivities provide the early history of the spectacular tradition of public anatomies—a tradition that reaches its fullest expression inside the (second) permanent anatomical theater (1594–1595).

# CHAPTER TWO

# Fabrici's Dominion

## The First Anatomical Theater

In Padua, the young scholar and anatomist Girolamo Fabrici took the chair of anatomy and surgery in 1565, three years after the death of Gabriele Falloppio.[1] On 18 December 1566, he began his first public anatomy demonstration, which lasted almost three weeks.[2] Georgio Palm, a student from Nuremberg and the president (*consiliarius*) of the transalpine students at the university, noted that Fabrici's demonstration was praised by local and foreign students alike.[3] Descriptions of this kind of approval were fairly typical. Earlier anatomists (Andreas Vesalius, Realdo Colombo, and Falloppio) drew attention to their admiring listeners and spectators and to audiences that were engaged and supportive, and students corroborated these descriptions in their notes.[4] In 1560, for example, the transalpine students described Falloppio's demonstration as "diligent and accurate," noting in particular his interest in their "contrary opinions" and his pace, where nothing was hurried.[5] Nevertheless, support for Fabrici was surprisingly fleeting.

From the late 1560s to the early 1580s, Fabrici regularly disappointed medical students. In 1568, only two years after his lauded debut, Fabrici gave a demonstration using one male cadaver, one female one, and a living animal [*pedicus vivae*]; the Silesian student, Lassaro Salio said that "if it was not the most accurate, at least it was tolerable."[6] Salio lavished praise not on Fabrici's demonstration or his dissection of the specimens, but on the "most honorable" burial of the remains, which marked the conclusion of the event.[7] Complaints were made again in 1572, when students described Fabrici's demonstration as unimpressive, and similarly in 1574–1575, when they said that he began late and finished early and that his demonstration was "brief" and "not very exact."[8] In the next decade, these expressions of dissatisfaction grew into protests made in 1582, 1584, 1586, and 1588.[9]

Although the situation was dire, the detailed complaints of medical students (recorded in the *acta* of the transalpine nation, the *Atti della nazione germanica artista*) attest to their nuanced appreciation of anatomical demonstrations. Their notes give particulars on the success and failure of anatomists, and these precise details indicate important distinctions and elisions between public and private demonstrations. On two notable occasions, Fabrici succeeded in delighting his audience, and, in both instances, students noted his "complete" and "diligent" dissection, criteria that his unsatisfactory demonstrations failed to meet.[10] In notes such as these, students isolate the pedagogical habits worthy of praise or blame, and they identify aspects of the audience's pleasure at and boredom with anatomical exercises. Though the history of anatomy has sometimes been told as a story of heroic achievement, the details of frustration and failure revealed in these sources help to reframe the history of anatomy and the anatomical theater by illuminating two features: the development of Fabrici's research and the development of a public academic tradition of demonstrating anatomy that was philosophically charged. While Fabrici's research has been seen as a product of his study of Aristotle and, in general, of the several "Aristotelianisms" on offer in Padua, Fabrici's turning to Aristotle distinguished his work from other anatomists.[11] As the students' records make clear, Fabrici connected his anatomical inquiry to the concerns of Aristotelian natural philosophy, in part because in doing so, he could distinguish himself as philosophically superior to his competitors. The element of competition, while certainly forceful in earlier decades, is crucial in the post-Vesalian era because it helps us to understand how and why the public demonstration of anatomy became philosophically, rather than visually, emphatic.

During this period, public anatomy demonstrations returned to a lecture-style format (discussed in chapter 1), but one intended to promote the philosophical dimensions of anatomical inquiry, that is, the principles that explained anatomical formations in a wide range of species. Though public anatomies have been described as gory—even violent—explorations of corporeal matter, Fabrici's demonstrations did not foreground dissection or the observation of anatomical structure, but rather the Galenic categories of *actio* and *utilitas*, or function and use (the latter underlining the natural philosophical dimensions of the topic).[12] The emphases struck the students as strange, and they began to compare Fabrici's approach—the content of his demonstrations and his style—with the private anatomical exercises that occurred in tandem with the public ones. Once again entangled with each other, the public and private venues of anatomy redefined the nature of the anatomical spectacle and its claims to a new kind of visual experience.

In increasingly clear terms, Fabrici's demonstrations come to contrast with private anatomical exercises. In the hospital of S. Francesco and in local pharmacies, private anatomies provided alternative anatomical instruction for often-dissatisfied medical students. These anatomies emphasized the anatomical causes of disease, aspects of morphology, and the technical procedures of dissection as well as the more intimate interaction between students, teachers, and cadavers. They also rivaled Fabrici's demonstrations. The students' preferences for private anatomies, along with a series of institutional reforms, intensified the competition between Fabrici and other anatomists. Not only did the immensely competitive environment motivate Fabrici to distinguish his anatomical program from that of his competitors and thereby emphasize the relationship between anatomy and natural philosophy in his public demonstrations, but it also led him to seek control over *all* of the *public* academic proceedings on anatomy and the spaces in which they occurred.

Fabrici, in solidifying the link between spaces of instruction and teachers, sought dominion over the public anatomy demonstration and the first permanent anatomical theater. This theater, a small structure, much like scaffolding, that was built in a corner of Palazzo del Bò between 1583 and 1584, was celebrated by medical students for its permanence. In the academic culture of Padua, where particular professors were connected to specific spaces of demonstration, the theater became an instrument of power. There, Fabrici dramatically staged the natural philosophical dimensions of anatomy and further distinguished the public event from the private lessons and exercises conducted by other medical professors in the local hospital and by other anatomists in local pharmacies. In the students' portraits of their teachers and their classroom activities, we see the public anatomy demonstration, the anatomy lecture, and the private exercise being reconfigured. These provide the historical conditions from which a new kind of anatomical spectacle emerged.

## ANATOMY AND NATURAL PHILOSOPHY

Between 1565, when Fabrici took his chair, and 1583–1584, when the first anatomical theater was built, Fabrici taught anatomy as a philosophical or natural philosophical inquiry. While medicine and natural philosophy shared a long textual history, and anatomists had sometimes taken their quests in this direction—Vesalius said that anatomy needed to be recognized as a branch of natural philosophy—Fabrici was more consistent and more committed to this intellectual project.[13] He treated the question of whether anatomy was a basis for universal knowledge (*scientia*) and emphasized ways in which anatomical inquiry engaged the intellect in the formation of principles, not practices. For Aristotle, the soul was the seat of two rational faculties, the *intellectus speculativus* and the *intellectus practicus*: the former speculates on unchanging entities, such as principles for the sake of truth, while the latter considers changeable ones for the sake of *opus*, or activity.[14] Emphasizing the *intellectus speculativus* and the ways in which anatomical knowledge qualified as *scientia*, Fabrici indicated that his students were supposed to learn the principles that explained the formation of similar anatomical structures in a variety of species. His intent was for students to contemplate the parallel anatomies of species, moving between particular features of anatomy and axiomatic and general explanations of those particulars; to understand a process such as digestion, respiration, or generation (the coordination of which required sets of anatomical structures and multiple functions); and to hold in the mind's eye an idea of the processes and principles of the "Whole Animal."[15]

Fabrici began to develop his approach, and this kind of inquiry, by focusing his demonstrations on specific topics. Between 8 January and 7 February 1575, for example, he gave a public anatomy demonstration and, somewhat surprised by his attention to detail, the transalpine students,

Johanne Exlino and Christophoro Sibenburger, mentioned that his dissections were done with "such care and such industry that even someone who watched only as a stone [would watch] would easily have received exact and complete knowledge of the composition of the human body."[16] This demonstration was "complete" in the sense that it covered the anatomy of the whole body; in other words, it was comprehensive, as public demonstrations were supposed to be. It was also atypical, for the students implied that Fabrici's demonstrations did not usually achieve either such coverage or such clarity. Indeed, in the previous year, they complained that his demonstration was "brief" and "not very exact."[17]

While many university lessons were suspended in 1576–1577 because of the plague, in the following year Fabrici organized his anatomical instruction more fully around discrete discussions of specific anatomical parts. In the winter of 1577–1578, Fabrici was unable to give the public anatomy demonstration because he was sick. At his home and for a smaller audience, however, he dissected pieces of a cadaver and exhibited them to spectators; two students, Christophoro Sibenburger and Samuele Keller, said that his treatment of these few pieces "consumed" him for days.[18] Fabrici's focus on a few pieces of the cadaver did not imply industrious dissection, because the students distinguished that phase of the demonstration as having already been completed. Rather, it was the treatment (tractationibus) that "consumed" him, or kept him busy. Unlike Falloppio, Fabrici did not tend to emphasize the cutting and showing of the specimens, the discovery of structures, or even training in observation. Fabrici talked. His treatment was not comprehensive, and it kept him from responding to his students. For both reasons, students noted that he attempted to "evade all things anatomical."[19] Fabrici may have discussed the relationship between the pieces of the cadaver, the role their anatomy played in a coordinated action such as digestion, and the natural philosophical dimensions evident in the foregoing discussion and in the commentary tradition. What is certain, though, is that this treatment consumed him and alienated his students to such a degree that despite the intimate domestic setting of this lesson, students found him distant and evasive.

For his demonstrations, Fabrici would choose topics and sets of structures. Based on Galen and on Fabrici's teachers and contemporaries, it was expected that a typical anatomical account would cover, first, *historia*, which was both dissection and the description of structure (including position, form, dimension, and construction); second, *actio* (or function), which referred to how the part functioned in relation to either the vegetative or the sensitive soul; and third, *utilitas* (or use), which referred to the final result or principle.[20] Fabrici began by limiting the number of significant structures only to those that were important to the topic. Placing limits at this early stage of a demonstration clearly diminishes the potential for discovering new structures. It suggests that the structures of any given topic were known, making their dissection less significant. Fabrici's publications indicate that he then looked at the specific set of structures in a range of animals and extrapolated principles based on their similarities and differences. Finding the general incidence of a part, he explained what was usually or normally the case in animals: that a structure has a certain shape, position, and substance.[21] Fabrici explained "use," the third category in Galen's scheme, with the example of the eye: the eye is an anatomy with many parts, but its main use is to transmit images "to the brain's faculty (imagination, reason, and memory) so that they may discern from these images what is true and false, beneficial or harmful, for the purpose of pursuing the one and avoiding or fleeing the other. This, finally, is an action which is useful to life itself, that is, to a still more comprehensive action, as Aristotle says."[22] Fabrici called this action a "public" action rather than a private one, and he made its moral valence explicit.

Fabrici's attention to particular processes or topics meant that his actual dissections did not cover the whole body, which was very unusual. Fabrici himself recognized this, such as in the beginning his work *On motion* (1618): "For this whole time in which we are administering not a popular anatomy, but an exact one, I decided, o listeners, with divine help favoring, and for your pleasure, to deal with the motion by which the whole animal moves with respect to location . . . the motion by which the whole animal changes its place or position."[23] Fabrici used "popular" to refer to comprehensive dissections and demonstrations, ones that covered the anatomy of the whole body and often served as introductions for medical students. This was what the public demonstrations typically achieved. In contrast, Fabrici called his demonstration, or administration, "exact" because it was topical, a treatment of the structures related to the topic of motion. It was still public, but he wished to flag it as different from the usual and more general content of public demonstrations. Unlike his predecessors and contemporaries, who satisfied students by demonstrating the techniques of dissection and offering a comprehensive, often structural account of the human body, Fabrici limited his treatment—even in the public anatomy demonstration. A dissection of the entire body was not part of his program, nor was the comprehensive description of human structural anatomy.

These topics, or processes, derived from Fabrici's study of Aristotle's works. As Andrew Cunningham has explained, similar to Aristotle's *oeconomy* of the Animal, Fabrici identified the Whole Animal as his object of inquiry and sought to provide a single, general, and universal account of a coordinated action or process.[24] These processes were the vital functions of the organic soul, which included motion, sensation, digestion, respiration, and generation.[25] Where Aristotle studied the rational soul, Fabrici produced an account of speech; where Aristotle studied the motive soul, Fabrici produced a work on locomotion; where Aristotle developed his ideas on the sensitive soul, Fabrici wrote on vision and hearing; and from Aristotle's studies of the vegetative soul, Fabrici published tracts on digestion, respiration, and generation.[26] Moreover, Fabrici, paralleling Aristotle, sought to produce causal knowledge: *scientia*, or an understanding of the first principles of the Animal. He thus directed attention to the relationship between anatomy and natural philosophy.

In his work on vision, *De visione, voce, auditu* (1600), Fabrici cites the classical sources appropriate to each stage of his investigation. For the *historia*, he recommends Aristotle's *History of Animals* and Galen's *On Anatomical Procedures*; for the study of actions, he recommends Aristotle's *On the Soul* and *On the Generation of Animals*, as well as Galen's *On Natural Faculties*; and for uses, he points to Aristotle's *On the Parts of Animals* and Galen's *On the Uses of the Parts*.[27] This approach also defined his pedagogy. In *The Formed Fetus* (1600), Fabrici explained his "customary order": "I shall first inspect the parts relating to this matter, in other words, make a dissection; then treat of action; and in the third place, inquire into the uses of all the parts, not the parts of the fetus, of course, but those parts which look toward this care."[28] Though Falloppio had emphasized his ability to cut and show and the importance of observing, Fabrici bracketed these as initial labors, necessary to distinguish typical from accidental structures, but not nearly as significant as capturing and making intelligible the first principles of anatomy.

Fabrici also did not elaborate on the technical procedures of dissection. To do so would be to emphasize that the study of anatomy engaged the *intellectus practicus* and was directed toward activity related to the medical and surgical ends of anatomy, and especially manual ones. Instead, he emphasized the movement from particulars to general axioms and universal principles and the necessity of a philosophically complete account of anatomy. Even when Fabrici made a structural discovery in the course of his demonstrations, he used it as an occasion to emphasize the philosophical nature of his inquiry. Take, for example, his famous discovery of the *ostiola*, or doors, of the veins in 1574.[29] These doors in the veins were a structural discovery that could easily be placed in the context of other such discoveries: Colombo's discovery of the clitoris, and Falloppio's discovery of the fallopian tubes. Fabrici expressed his surprise at seeing the *ostiola*, but he then described the discovery as a "cause for wonder."[30] Aristotle had noted that philosophy (as the search for final causes) began in wonder; Fabrici echoed him and simultaneously redirected the attention of his audience to the search for causes rather than solely to the isolation and appreciation of structure. As to why earlier anatomists missed these membranous doors, he wrote: "Either they neglected to investigate the function of the doors, a matter, one would think, of primary importance, or else they failed to see them in their actual demonstration of veins. For in the bare veins exposed to view, but still uninjured, the valves in a manner display themselves."[31] It was not so much inattention or careless observation that confounded earlier anatomists—the doors were utterly visible, "displaying themselves." Rather, for those who missed the doors, the mistake was a conceptual one. They neglected to investigate the function ("actions") of the veins. Fabrici explained his discovery in relation to the coordinated action of respiration, responding to the limitations of anatomical inquiries that pursued structural composition over a full-bodied philosophical account.

## METHOD

Fabrici's method of inquiry had consequences for the rhetorical presentation of anatomy. As chapter 1 explained, Falloppio's methodology deployed the trope of revelation—he revealed structures—but it was grounded in the accumulation of experience through dissection and through trained, diligent observation. In contrast, Fabrici's rhetoric was filled with wonder, with coordinated actions as objects for contemplation, and with self-revelatory corpses. Falloppio wrestled with his specimens; but for

Fabrici, as he said of the *ostiola*, "in a manner, [they] display themselves." In other words, the body became more transparent and less an agent of obstruction. This rhetorical displacement had its correlate in the anatomist, who was recognized as a philosopher (and thus distant from the procedures of dissection). The task of the anatomist was to search for the secrets and mysteries of nature, but these were not merely the fragile structures thought to reside in the innermost recesses of the human body (as a cursory glance at Vesalius's title page might lead one to believe; see figure 4 in chapter 1).

An anatomist was supposed to formalize the connections between animals in terms of their parts, functions, and purposes. This kind of inquiry promoted eloquence and contemplation, features already written into Galen's category of use. Fabrici claimed, for example, that his study of generation allowed anyone "to understand and to contemplate those first beginnings of the life of every animal."[32] Although these first beginnings, or principles, assumed that one had an understanding of structural composition and morphology, they were not things that could be seen in a decaying corpse or pulled out of its cavities. The focus was not on the industry and toil of the anatomist or on his skilled hands that performed delicate dissections. Rather, from self-displaying corpses, Fabrici could illuminate the purposive causes of anatomy, provide explanations for why things existed as they did, and thus demonstrate anatomy as a natural philosophical enterprise. This emphasis might also be read in Fabrici's subsequent publication project, *The Theater of the Whole Animal Fabric*, which contained an expansive series of colored pictures.[33] These images represented the structural characteristics of anatomy in a range of specimens and depicted parts located within an animal's body. While the illustrations served a range of functions, it may be that, given Fabrici's focus on principles, they were substitutes for lengthier, more detailed accounts of structure.

Fabrici's approach was rhetorically and conceptually innovative. Moreover, it was closely connected to the treatise on anatomical method that Girolamo Capivacci (1523–1589) drafted in the 1580s. In his *De methodo anatomica*, Capivacci, also a professor of medicine in Padua, sought to assimilate Aristotelian terms and concepts with Galenic anatomy.[34] Echoing Galen, Capivacci explained that anatomy is not only "to be considered by way of dissection, but also truly by way of actions and uses."[35] Elsewhere, he employed the Aristotelian terminology of causes, emphasizing the necessity of moving from material to formal and final causes. He argued that anatomical inquiry, pursued in this way, could produce philosophical knowledge (*scientia*). This, Capivacci declared, "is to be distinguished from the domain of skill [*peritia*], which is pertinent to dissections and to manual activities [*manuales*]."[36] *Peritia* was used here to refer to manual skill, the techniques that were shared by anatomists, surgeons, and perhaps some barbers. Although other writers would extend this term to manual expertise, Capivacci severely limited it to an instrument of knowledge (not a harbinger of expertise). For Fabrici and Capivacci, the anatomist did not leave behind the manual activities associated with dissection. Instead, a knowledge of causes could be derived from the description of a part, and the anatomist (along with his students) is celebrated for his ability to explain why a thing exists, rather than merely to describe it.[37]

This was the method that underlay Fabrici's anatomical inquiries, a method highlighting the relationship between anatomy and natural philosophy. Fabrici did not turn to natural philosophy to ask questions about the immortality of the soul, or about the unity of the intellect or free will; the academic community had achieved some consensus on these topics by the end of the sixteenth century.[38] Rather, what drew him was the allure of the systematic. Following such comprehensive texts as Vesalius's *Fabrica*, Fabrici proposed a project of inquiry that culminated in an encyclopedic account of the Whole Animal, organized by principles. The long history of natural philosophy made it receptive, at this point, to projects of encyclopedic dimensions, which were seen as significant and contributing to knowledge, rather than limiting access to nature.[39]

In addition to its encyclopedic arc, Fabrici's inquiry presented itself as a response to questions of medical authority and expertise. Anatomy was important to this academic culture, not because it made surgeons or physicians more effective practitioners, but rather because both physicians and natural philosophers could participate in the study of nature. Capivacci constructed an academic medical hierarchy with surgeons at the bottom and philosophers, rather than physicians, at the top.[40] Although it may seem as if he reproduced the age-old division between learned and popular medicine, between professors and practitioners, and, finally, between intellectual and manual labor, Capivacci instead attempted to construe and stabilize that boundary in a way that cast Fabrici's work in the field of anatomy on the side of philosophy and methodological innovation.

Fabrici's studies would be given a more secure place in the institution if they were accompanied by a clear method such as Capivacci's.

It was also the case that Capivacci's discussion of anatomical method appeared in the wake of considerable pedagogical confusion. In this, anatomy mirrored other disciplines. Anthony Grafton and Lisa Jardine have explained that the recovery of ancient knowledge quickly produced problems in the classroom—variant approaches to the texts, a lack of curricular unity, and the availability of only certain texts—that were eventually subsumed under the discussion and practice of method.[41] Although their analysis pertains to the late humanist traditions of teaching and scholarship in the language arts, a similar situation characterized the anatomical tradition.[42] This was one reason why Capivacci felt the need to dedicate an entire monograph to the topic of anatomical method and to assimilate the two frameworks of explanation, melding Galen's threefold system with Aristotle's fourfold system of cause. A second reason for Capivacci's interest in method was the debates around *ordo* in philosophy that engaged Francesco Piccolomini (1520–1604) and Jacopo Zabarella (1533–1589), among others. As Nicholas Jardine has explained, both Piccolimini and Zabarella considered philosophy to be devoted to the rediscovery of lost knowledge from the philosophical writings of Greek antiquity, but Piccolimini included metaphysics, along with the study of nature and the theoretical part of ethics and politics (*scientia civilis*), while Zabarella restricted it to contemplative disciplines, distinct from ethics and medicine.[43] Fabrici was closer to Zabarella than Piccolomini. Fabrici treated the more philosophical texts and aspects of both Galenic medicine and Aristotelian philosophy, effectively raising the status of anatomy by developing its philosophical and contemplative dimensions. For this reason, he considered his appointment to the ordinary chair of anatomy in *primo loco* (the first rank) in medicine, which was made in 1584, to be entirely appropriate, because anatomy had a more developed philosophical foundation.

In the second half of the sixteenth century, moreover, there was a broad reconsideration (if not a restructuring) of the disciplines of knowledge at the university. After the war of the League of Cambrai (1509–1517), Venetian officials made an effort to hire professors for Padua from other institutions and from foreign locations, and this allowed intellectual practices associated with medical humanism to prosper.[44] The climate also supported an intellectual exchange between professors of practical and theoretical medicine. In a letter to Ulisse Aldrovandi, Falloppio explained: "These men are all about changing the [medical] school, putting courses from practical medicine into theoretical medicine and theoretical courses into those in the practical branch, and I believe that without doubt, the chair in practical medicine will stay empty [as a consequence of Vettor] Trincavilla and [Antonio] Francanzano going over to Theory."[45] The boundary between the medical branches was permeable, allowing for the movement of professors from one to the other and especially from the practical to the theoretical. Subsequently, however, sharper lines were drawn between fields and disciplines. Chairs did not revolve as easily. Fabrici himself gained the extraordinary chair in anatomy in 1565 and the ordinary chair in 1584, working against Giulio Casseri in the early decades of the seventeenth century to maintain it. By emphasizing the philosophical dimensions of anatomical inquiry as part of its method, Fabrici and Capivacci sought to relocate anatomy, to bind it to the theoretical branch of medicine and to traditions of natural philosophy that were seen as formidable, laudable, and unquestionably legitimate.[46]

Capivacci must have considered his students carefully, for medical students continued to protest the lack of practical medical and surgical instruction in anatomy demonstrations. The students registered their disappointment with Fabrici's demonstrations in precisely these terms. Not only did they look elsewhere for anatomical instruction, but they also began to judge Fabrici's demonstrations against alternative, private anatomical exercises. An early instance of this came in 1572. Reminded by administrators of his teaching responsibilities, Fabrici held a private dissection, which the transalpine students Girolamo Fabro and Daniele Fuschsio described as not altogether "unfruitful."[47] Perhaps looking for a more detailed treatment, the students asked Nicolò Bucella (ca. 1520–1599), then a surgeon and "most learned" in the "anatomical business," to consider giving a public anatomy, since one had not taken place that year. Bucella was known to offer anatomies from his house in the S. Maria dei Servi quarter in Padua, and the students seemed sure that Bucella's demonstration would be nothing short of splendid, but, in the end, cadavers were lacking.[48] Students continued to seek additional anatomical exercises: in the 1570s from Bucella, and in the late 1570s and 1580s from Paolo Galeotto and Giulio Casseri. Students judged one anatomy demonstration against another, one set of theatrical and pedagogical features against another, and one demonstrator against another. As always, anatomical proceedings depended on the availability of material. Capivacci sought to lay out in the clearest terms possible what the point of Fabrici's anatomies was, in order to provide a method that would convince students, in addition to professors, of the merits of Fabrici's instruction.

## PRIVATE ANATOMIES

Students found sources for alternative instruction on anatomy from private anatomical exercises (which I will discuss in a moment) and from medical classes that took them into the hospital. During clinical rounds at the hospital of S. Francesco, professors taught students how to treat patients, take a pulse, analyze urine, and recommend appropriate courses of treatment.[49] During the 1570s and 1580s, these private courses were taught by professors of both theoretical and practical medicine: in addition to Capivacci, others were Girolamo Mercuriale (1530–1606), Albertino Bottoni (d. 1596), Marco degli Oddi (1526–1591), and Emilio Campolongo (1550–1604). During these decades, the transalpine students repeatedly praised these professors for their learning, their pedagogical skill, and, with respect to their forays into anatomy, their practical orientation.

The practical orientation of these lessons derived not only from the professors' interest in the medical uses of anatomy (rather than the natural philosophical ones) but also from their location in the hospital of S. Francesco. In 1578, when the weather turned cold, both Bottoni and degli Oddi decided, since they were holding lessons in the hospital, to dissect the bodies of two women who had recently died. The decision to open these cadavers was made in order "to demonstrate to the listeners the affected places and the kindler [or cause] of the diseases [fomites morborum]."[50] They wished to dissect the uterus in each cadaver. Then Campolongo, a professor of theoretical medicine, joined the group and said that in the second cadaver, the body of a woman "consumed by senility [marasmo]," he would "penetrate the fistula" that resided beneath her breast, cutting into it in order to show its parts and connections more clearly.[51] The examination was then interrupted by the complaints of a little old woman (quaerelis anicularum), who probably feared a similar treatment, exposure, and delay of burial at her own death.[52] The university threatened degli Oddi and Campolongo with the loss of their salaries; and at the beginning of the subsequent academic year, the riformatori reminded everyone of the statutes governing anatomical proceedings. Although bodies for dissection were supposed to be both ignoble (criminal) and foreign, the *riformatori* instead emphasized the constraint on time: the beginning of the anatomy demonstration had to follow the end of the anatomy lecture, and everything was to occur in the break between Nativity and Carnival, when students were not in class.[53]

Although it is unclear if the anatomies of these two women were finally conducted, they were intended to instruct students on more practical aspects of medicine—the causes of disease and the particular anatomical traits (of the uterus and breast) of the female body that could be read as signs to discover the nature of the diseased wombs and the observed fistula.[54] The setting encouraged this practical orientation and, equally, the improvised nature of the scene. The dissections were intended to be medical anatomies. As Campolongo indicated, his dissection would penetrate the body and reveal the characteristics and causes of the diseases lodged within. Oriented by medical topics—the nature and cause of disease—anatomies such as these were occasional, improvisational, and underregulated.

Meanwhile, degli Oddi, Campolongo, and the other professors promoted the practical aspects of medicine in their teaching. During the 1570s, when plague struck Padua and Venice in successive waves, they taught diagnostic skills in their courses, as well as the history and contemporary understanding of plague and contagion.[55] In 1576–1577, for example, Mercuriale lectured on the diseases of the "rational faculty" and the brain before making a brief dissection and then finishing his commentary with a discussion of the plague, "as the *riformatori* requested."[56] Subsequently, an alternating sequence of instruction began: in the morning Mercuriale explicated the Hippocratic work on epidemics, and in the afternoon Capivacci lectured on the methods of curing disease, a subject he "revealed and explored thoroughly [*retexuit et pertexuit*]."[57]

In contrast to the timely focus of Mercuriale's and Capivacci's exercises and the practical (though underregulated) lessons of Oddi and Campolongo, in the winter of 1578 Fabrici addressed his students in a public lecture and a private colloquium, "as he had promised."[58] The students, Samuele Keller and Ioanne Wolfango Rabus, said Fabrici "filled and elevated their souls with many things."[59] Fabrici inspired his students not because he was dissecting corpses and vivisecting animals, but rather because he connected anatomy and natural philosophy, studying the coordinated processes of the soul, that is, the life force responsible for motion, digestion, respiration, sensation, and generation.[60] Fabrici's lectures and demonstrations were thus clearly distinguished on their own merits, and in the minds of students, from the medical anatomies held in the hospital of S. Francesco.

While Fabrici began to pursue the relationship between anatomy and natural philosophy, the public anatomy demonstration continued to be the site of formalized debate. Disputations were an important form of exchange and learning at the Renaissance university. During a formal disputation, one party would defend a thesis against a second or a third party's repeated attacks. In Fabrici's case, he would defend his claims about anatomy. In 1579–1580, for example, Fabrici first equivocated about giving an anatomy demonstration. In December, he said he would give a private anatomy, which students described as "inadequate [admodum exigua]," and one month later he agreed to give the public anatomy demonstration.[61] This time, the rectors of the city had shown up. When, after "a lot of rattling and whistling," everything turned "tranquil" (on 17 January), Fabrici began his demonstration, dissecting "with singular industry" male and female cadavers and vivisecting a cow.[62] Following this work, a debate ensued. The student writing the report noted that professors raised their doubts before noon, and these were discussed and resolved in the afternoon.[63] Although this anatomy demonstration began with the dissection of several cadavers and the vivisection of an animal, it was devoted to Fabrici's presentation and his responses to the professors in the audience.

For the students, however, the demonstration was noteworthy for two other reasons. First, the transalpine student who described the event, Francisco Hippolyti Hildesheim, highlighted Fabrici's industry and the initial dissection (consectis singulari industria), suggesting that this stage of the demonstration was of great interest to medical students. Secondly, he contrasted the participation of professors in the demonstration, who interacted directly with Fabrici and the points he presented, to the students' subsequent participation. At the end of the demonstration, on 6 February 1579, Hildesheim explained that "the remaining parts of the cadavers [were] carried to the Cathedral and buried" and that the burial was attended "not by many professors, but by many students, among whom there was the greatest number of us [the transalpine students]."[64] While the students' interest in the burial ceremony is discussed more fully in chapter 4, here it highlights an important contrast. In the public demonstration, professors, rather than students, disputed the points of anatomy; thus professors, rather than students, were the ones who participated fully in the anatomical event.

## REFORMS

Students, however, wanted to take part in more aspects of the annual event than the burial ceremony that concluded it. They also wanted more opportunities to study anatomy, a desire expressed in several of their complaints and in their decision to contact university administrators. In 1581, Fabrici promised to give "a most complete and highly illuminating" anatomy demonstration that focused on tumors and fractures; though patient while waiting for the anatomical proceedings to begin, the president of the transalpine students, Christophoro Haenzelio, found Fabrici's demonstration to be "very obscure and very imperfect."[65] There were more complaints in 1582–1583, because Fabrici neglected to give the public demonstration and, as students said, he was demonstrating every other year.[66] In January 1582, the transalpine students contacted the *riformatori* in order to explain Fabrici's neglect and express their desire for more opportunities to study anatomy.[67] Because it was a part of the "expectations of their studies," they wished the higher authorities "to compel Fabrici to give a public anatomy every year" or to have private anatomy lessons in the years that Fabrici did not offer public demonstrations.[68] By 1582, and given the preceding history of complaints, these students were motivated to make their petition not only because Fabrici failed to demonstrate annually, but also because when he did, they remained dissatisfied.

In addition, the students were upset since, when they organized a private dissection, they often could not obtain a cadaver. On 19 January 1582, the students organized and prepared for a private dissection, which would be given by the surgeon Michael Aloisius in the church of St. Catherine. The scheduled event was not officially approved, and although the students convened and discussed the supply of instruments and bodies, the private dissection did not take place: "because of scarcity, we were able to acquire no cadaver . . . our deliberations were carried out in vain."[69] The scarcity of cadavers is nearly universal in the early history of anatomy (until the end of the century), but here it served to increase the disappointment and frustration that the students felt toward Fabrici and to strengthen their desire for supplemental, private anatomies.

In 1583, the *riformatori* began a series of reforms that further clarified Fabrici's responsibilities and separated the content of private anatomical exercises from that of the public demonstration.[70] Noting the "utility and great honor" that Fabrici's work brought to the university, the *riformatori* decided that Fabrici would conduct the anatomy—as he had done "year after year"—and also lessons on surgery. Both would be held in the winter months because it was impossible to "manage the corpses in the warm months"; and, since Fabrici would be conducting the two kinds of lessons, he would not need to interrupt the ones on anatomy with surgical lessons, because the lessons on surgery would be separate.[71] Fabrici received 400 florins for his anatomy demonstrations and, for his surgical demonstrations, another 200 florins.[72] To put this in some perspective, in the early sixteenth century, a skilled craftsman at the Arsenal in Venice would earn around 50 florins a year; a communal *podestà*, 190 florins; a professor, between 100 and 600 florins; a shipmaster, 720 florins; and a Doge, 3000 florins.[73] Even with the inflation of the florin by the end of the sixteenth century, this raise and those that followed—Fabrici's income from his academic work eventually topped 1100 florins—were dramatic, a sign of the university's commitment to Fabrici and its recognition of the growing importance of the field of anatomy to both medicine and philosophy and to the reputation of the institution at home and abroad.[74]

Although the reforms assigned instruction on both anatomy and surgery to Fabrici, he only offered surgery demonstrations infrequently. The surgical uses of anatomy, as we will see in a moment, were often discussed in the private anatomies given by Casseri and Galeotto. This distinction between anatomy and surgery indicates the growing independence of anatomy as a field of inquiry. Historically, the chair of anatomy and surgery was held by a practicing surgeon, who was responsible for the anatomy demonstration as well as a three-year cycle of commentary on surgical works.[75] In Padua, the statutes provided for two chairs, the ordinary one, which was more prestigious, and the extraordinary one, which was less so. The salaries for these chairs were low, and the statutes allowed one person to occupy both, at which point he would receive both salaries. The professor holding one or both of these positions also usually worked as a surgeon in a hospital during portions of his tenure. Fabrici did not work at the hospital, but he maintained a lucrative surgical practice on the side.[76] Casseri, who sometimes served as a lecturer in surgery and who held the chair of anatomy and surgery after Fabrici, continued to work at the hospital of S. Francesco. With its 1583 decision, the university and the *riformatori* encouraged the perception that anatomy, not surgery, was the field under development.

The decrees also differentiated the content of the annual anatomy demonstration from surgery demonstrations. Previously, the same anatomist would move between public and private venues. In both venues, the anatomist could highlight the (natural philosophical) principles and purposive causes of anatomy and then emphasize the importance of knowing anatomy in order to understand (medical) sources of disease and the (surgical) ways to set fractures. The administration's decision distributed these functions among faculty members: Fabrici gave public anatomy demonstrations that were more philosophical in character; and Casseri and Galeotto administered private anatomies, emphasizing the medical and surgical uses of anatomy. In 1585–1586, Casseri offered a private anatomy, and the students said that "by dissecting very carefully all the parts of the body—not only the internal parts but also the external (i.e., the muscles and veins)—and in addition by demonstrating the main surgical operations," Casseri "acquitted himself excellently and satisfied us all."[77] Dissecting skill and the areas of the body that displayed that skill (such as the muscles and veins) were routinely connected to more practical considerations and to the topic of surgery. Although the distinction between internal and external treatment has been used to distinguish the medical realms of the physician (internal remedy) and the surgeon (external treatments), it doesn't appear to have had much force in this demonstration. Attention to the muscles would be related to fractures (for the surgeon) and to tumors (for both the surgeon and the physician); and both kinds of practitioners would benefit from attention to the arteries, veins, and nerves, which would be connected to bloodletting (recommended by physicians), cauterizing ulcers and other sores, and both cauterizing and stitching wounds. More consistently than Fabrici, both Casseri and Galeotto employed a pedagogical approach that relied heavily on the dissection of the internal and external parts of the body, on the display of surgical operations, and on the participation of students (a feature made possible by the more confined setting of the private lesson). Thus, since private lessons focused on manual techniques such as surgical maneuvers (in addition to anatomical particulars related to disease), the public anatomy demonstration could focus on the natural philosophical dimensions of anatomy, demanding a new conceptual sophistication from the students.

As chapter 1 indicated, Falloppio's public anatomy demonstrations were oriented by structure and by modes of apprehension related to both touch and sight. During Fabrici's tenure, these features migrated to the private lessons on anatomy. In contrast, the public demonstration began to focus increasingly on the philosophical dimensions of anatomy. This component had been present in the earliest studies of anatomy, but it was treated only briefly at the outset of a demonstration, to provide a suitable framework for the subsequent dissection; it was also explored more fully in lectures, which preceded the demonstration. Indeed, Falloppio separated the dissection and study of structure from the speculative treatment of anatomy. With Fabrici, however, such speculation was not relegated to the lecture; speculation was incorporated into the demonstration itself. Attention to sensory apprehension—manual practice as well as touching and seeing the anatomized body—was more thoroughly a part of medical anatomies and private dissections. Though medical anatomies occurred infrequently or were rarely documented, private dissection, thanks to the students, was in a state of resurgence.

## THE FIRST ANATOMICAL THEATER

When the first permanent anatomical theater was constructed, the public anatomy demonstration was run more like a lecture; this further distinguished it from the private exercise. Inside the permanent theater, the demonstration developed its philosophical content and diminished its procedural connection to dissection. Conceived as a part of the reforms of 1582–1583, and based on the political and financial support of Lorenzo Massa (the nephew of Niccolò Massa), the first permanent anatomical theater in Padua was completed by 9 January 1584.[78] The theater was described as a *theatrum publicum et perpetuum*, literally "a place for seeing that was public and permanent."[79] The theater was called "public" because it was intended to hold the public anatomy demonstration and to include an audience of students, professors, administrators, and, occasionally, magistrates. In only one instance did the first theater entertain a more diverse audience; in 1588–1589, students referred to the friends of the *massarii* and common people in the audience.[80] No other record of a more diverse community attending the demonstration exists until the opening of the second permanent anatomical theater in 1595, a documentary feature that suggests that even inside the first permanent theater, the annual demonstration was an almost exclusively academic event.

This theater was notable, however, because it was permanent. While temporary theaters were taken down each year and reassembled in the next, this theater was left standing. Most likely a wooden structure with stable scaffolding, it was built on the first floor in the corner room of Palazzo del Bò, the hub of the university in the sixteenth century.[81] When the students lauded the theater as "our new theater of magnificent Venice," they celebrated the financial and symbolic support that the study of anatomy received from the Venetian Senate. They also, and perhaps more importantly, celebrated the permanence of the theater, for, with a permanent theater, Fabrici would be inclined to provide rather than postpone the annual anatomy demonstration.[82]

The permanent theater was a place for seeing, but this does not mean that it enhanced the observation of anatomical material or emphasized visual apprehension. Rather, it resembled other academic theaters as a place for students and faculty to assemble, to hear lectures, to attend demonstrations, and, in general, to engage in academic instruction. In 1586, for example, the transalpine students mention an inaugural ceremony that took place in a "theater of letters." In this "cathedral building," a reverend monk spoke "elegantly and copiously" about "attaining the perfection of man and the intellect by letters and labors." He urged students "not to spare their labors" and praised "those who were supported in this place ['the most ample theater of letters'] by the munificence of the Most Serene Lord": "the reverend theologians," "the very experienced physicians," and "the very wise philosophers."[83] Hosting lectures, demonstrations, disputations, and inaugural ceremonies, this theater of letters served the academic community just as the anatomical theater probably did. Thus "public" referred both to the academic community and to the demonstrations—held publicly—for that audience. In the 1580s, instruction on anatomy (as well as in other fields of study) included lectures, demonstrations, and disputations, all of which were overwhelmingly auditory, not visual events. Moreover, anatomical instruction was not popular (in the sense that it intersected with a wider demographic), even though its events took place within ample and splendid theaters.

As a part of the reforms of the 1580s, the anatomical theater was first characterized as a structure that would regulate students. Although matriculation levels were steadily rising, the increase in the number of students studying medicine does not appear to have been drastic enough to warrant a separate, larger location for the sole purpose of hosting the annual anatomy demonstration.[84] Issues of space developed after this theater was in use, suggesting its success. In 1583, the administration decreed "that a public and permanent theater be built in the best lecture hall at the expense of the Venetian State" so that the anatomist could conduct his exercises "without any impediments."[85] The anatomist's impediments were partly financial. The organization of the annual demonstration was costly and complicated, and, typically at the end of the demonstration, the students in the audience would pay the anatomist for the temporary theater and scaffolding and for the burial of cadavers.[86] On 15 July 1583, when the Venetian Senate allocated 130 *lire* and 16 *denari* to the beadle for the conservation of this theater, it attempted to relieve Fabrici and the medical students of this financial burden.[87] This stipend was roughly equal to 22 ducats, or the annual salary of a sailor or a soldier in the early sixteenth century. Moreover, if students were not excited about going to Fabrici's demonstration and paying the additional cost, this money would have encouraged their attendance.

The term "impediments" also highlighted the students' frequent interruptions during anatomy demonstrations. At Vesalius's demonstration in Bologna in 1540, the Silesian student Baldassar Heseler described how Italian students mangled a vivisected dog, one example among many that attests to the obstacles that these disruptions posed to the progress of demonstrations.[88] During the same series, Vesalius rushed to end of his dissection of the head "because he was very confused, upset, and bewildered owing to some noise and disorder that the students then made [and] . . . being upset as he was very choleric, he hurried on this dissection."[89] Such disruptions also took place in Padua. In 1582–1583, some students in Padua stole a cadaver, which was designated for dissection on the following day; in the middle of the night, they mutilated it in an "indecorous fashion" and later disposed of it in the Brent River. In 1586, Fabrici's demonstration was suspended, due to the fact that the cadaver as well as the instruments had been stolen from the site of the demonstration.[90] Fabrici's thwarted demonstrations may be inflected with particular significance, for, in both cases, the entire specimen was destroyed and Fabrici's demonstration was brought to a grinding halt. Given that students criticized Fabrici for giving useless demonstrations, that is, for demonstrating inaccurately and tediously, perhaps they stole the cadaver and the instruments precisely because Fabrici seemed capable of demonstrating without them.

Historians and literary scholars have connected these kinds of disruptions to Carnival or to the carnivalesque atmosphere of turning authority on its head, but the misbehavior of students has a long history in the life of the Renaissance university.[91] Marc-Antione Muret, the eminent professor of rhetoric at the university in Rome, requested that he be relieved of his academic post for several reasons, one being the "perpetual insolence of the students"; in addition to noting that too many students carried daggers, he said he was forced to end a Saturday lecture when a student threw a melon at him "with the manifest peril of putting out an eye."[92] While misbehavior could be instigated by Carnival antics or intensified during the Carnival season, it was a widespread phenomenon that, in the sixteenth century, was on the rise. Student violence escalated; students perpetrated crimes against each other and against their teachers. Given this environment, the decree's reference to impediments probably also referred to the bad behavior that was typical (or typically expected) of students.

The decree cast the theater in a disciplinary role, lending its practiced formality a disciplinary function. Descriptions of assemblies at the university make it clear that formality was thought to translate into a more seamless, less disruptive, and thus more successful event. In November 1577, the students assembled for the annual nominations of professors in the ecclesiastical cathedral. There was "a most elegant oration in praise of our school" and "the professors were nominated and as is the custom, recited publicly."[93] The formality and peacefulness of this assembly were in direct contrast to the lectures and disputations that occurred only a few days later, when professors attempted to continue their "scholastic works [operas scholasticas]" but were repeatedly disrupted because of disagreements over the prohibition against weapons, which students were not allowed to bring to class.[94] Situated between these two extremes, the first permanent anatomical theater was conceived of as a mechanism for control rather than as a place for carnivalesque spectacle. This theater also helped to formalize the anatomical event and to remove disruptions and interruptions, which were the real impediments to a student's education.[95] Just as in other university assemblies—such as convocation, voting, and public examinations—students at the public anatomy demonstration were supposed to listen quietly to the rector (an elected student position) and to the speakers on the proposed subjects; those not wishing to stand or sit quietly were to be excluded and, if further trouble ensued, the student would lose his voting privileges for four months.[96] The rector was supposed to enter first, taking the first bench, and to be followed by the doctors or professors of the university. No student could sit in the first row without incurring a fine of 20 *soldi*, unless the beadle granted him permission.[97] The fine of 20 *soldi* (or roughly a *lira di piccoli*) was approximately the cost of entry, though that fee ranged from as little as three small silver coins (*marcelli d'argenti*) to 32 *soldi* in 1588. Students, moreover, were supposed to have studied for at least one year before attending an anatomy demonstration. Such practices and procedures existed for the earlier, temporary theaters, but the permanence of the first anatomy theater brought a new emphasis to them and to the twinned themes of discipline and regulation.

In addition, a second decree was issued that brought more clarity to the different kinds of anatomical instruction. It stated that before the anatomy demonstration, students had to attend preparatory lectures, where certain doctrines would be explained to ready them for "the contemplation of anatomy."[98] While these lectures may have also taken place in the new theater, the decree organizes anatomical instruction, something that was necessary when anatomy was being practiced and studied in a wide range of venues and among various professors as well as students. Demonstrations would include the dissection of cadavers and the vivisection of animals, while lectures would treat the more conceptual, philosophical dimensions of anatomy as objects of contemplation.[99] It was thought that both the preparatory lectures, which emphasized the philosophical importance of anatomy, and the theater itself would help to resolve the problem of student disruption. For Fabrici, the protocol for the public anatomy demonstration, plus these decrees, made the natural philosophical framework of anatomy transparent and expected. As Fabrici conducted his lessons, he began to emphasize the natural philosophical dimensions of anatomy inside the theater, in order to curtail disruptions and, at the same time, establish greater continuity between the demonstration and lecture, on the one hand, and his teaching and research, on the other.

The first permanent anatomical theater provided not only a material space that was permanent, but also a formal arrangement that displayed the hierarchy of academic posts. As the transalpine students explained, "in the new theater, the first seats had been assigned to the rector and the professors, and the next seats to the *consiliarii* [presidents of the student nations]."[100] While literary scholars and historians have taken *theatrum* to mean a place for seeing (as it does in Latin) and to imply a more concentrated visual focus on the dissected corpse, the students' description of the seating arrangement suggests two things. First, the audience was on display; while this is not unique to the anatomical theater, it is a notable development in the history of anatomy. Second, the spectators closest to the dissected material were the rector and professors, not the students. Fabrici's theater was a place to see the academic community assembled. It was also a place to hear Fabrici and to learn (and see demonstrated) the history of anatomy: its origins in classical texts, its traditions of commentary, and Fabrici's intervention in that history. Alongside the cadavers and the animals, Fabrici displayed his knowledge and his professional commitments.

Although the theater and the changes in the educational structure for the study of anatomy were meant to regulate the audience, such measures failed. In the same year (1584), Fabrici's fourteen-day anatomy demonstration was "disturbed by the rowdiness of the few Italian students (for it was their way)," so that many things fell from the anatomist's thoughts "that he otherwise would have set out to be discussed, to the great benefit of the audience."[101] Although the Italian students interrupted the event, the theater continued to be interactive, a space for professors in the audience to debate the points set forth by Fabrici: "furthermore, frequent objections were raised by the professors and discussed from both sides." While the degree of interaction between professors and the anatomist remained high—Colombo and Falloppio describe their interactions with other professors and their refutation of criticism in public demonstrations—the students' interaction was quite limited. Indeed, as they had in 1579, the students implied a contrast between the debates among the professors and the disruptions of the Italian students, with the latter emerging as the only real form of student participation.

In praising private demonstrations, however, students revealed their desire not for the debates that ensued between professors, but rather for their own participation. When their participation took appropriate forms, it included the ability to see the anatomical parts, to hear the structure and function of these parts clarified, and to enjoy a demonstration given by an anatomist attentive to his young listeners. When students described the success of private anatomy lessons, they did so in terms of what the public demonstration increasingly came to lack. Private settings offered instruction (occurring more frequently throughout the academic year) on structural anatomy, where medical students could interact with the teacher and the specimens as they had, for example, in Falloppio's demonstrations and on the hospital rounds with Marco degli Oddi and Emilio Campolongo.

## COMPETING LOCATIONS AND PROFESSORS

While the first permanent anatomical theater simplified the organization of the annual demonstration—there now were requisite preparatory lectures, permanent seating arrangements, and financial assistance for the anatomist—it discouraged the students from participating. Even when the permanent theater was in use, students commented more frequently on the other locations for lessons on anatomy and on the other teachers of anatomy. Between 1584 and 1592, when there is mention of the theater's destruction, the first permanent theater was one place among several that held anatomy demonstrations. As was repeatedly discussed by the transalpine students, Paolo Galeotto held anatomical demonstrations in at least two other locations, the church of St. Catherine and a pharmacy; in these locations, temporary theaters were constructed at the students' expense.[102] These locations and Galeotto's demonstrations presented so significant a challenge to Fabrici that he quickly pursued their prohibition, indicating that the link between teachers and the spaces of demonstration was gaining stability.

Whether in public demonstrations or in private lessons, students wished to learn structure-oriented anatomy within an interactive environment. While earlier students praised anatomists such as Vesalius, who demonstrated structural features and allowed the students to participate in points of anatomy that interested him, later students continued to prefer teachers who explicated points of anatomy clearly and, as the students themselves put it, responded to their particular needs. In 1584–1585, Fabrici gave a public anatomy demonstration that omitted the dissection of living animals (vivisection), which the students "had frequently desired."[103] In January of the next academic year, 1585–1586, the students appreciated Fabrici's public demonstration; he showed particular diligence in the demonstration of the liver, the spleen, the veins, the muscles with the skin removed, the uterus, "and other things not unpleasant to see."[104] Here—perhaps because he was more inclined to appease his students since he had received another raise of 250 florins on 28 January 1584, or because the corpses were well preserved—Fabrici had uncharacteristically focused on the structural anatomy of the exterior and much of the interior of the body, though he concluded the demonstration with the organs of generation or, more specifically, with the process of generation, a topic important to his research.[105]

Fabrici was also aware of the competition, which may explain his uncharacteristically thorough demonstration. In the same year, another cadaver was obtained for the private lessons of Giulio Casseri, who dissected "all the parts of the body" and provided instruction on surgery.[106] It met with praise, as did the private anatomy that Galeotto gave; students commended the latter for its "special usefulness" and for its adherence to the Galenic method.[107] These endorsements indicate what we might expect, namely, that students wished to learn the position, structure, and function of internal organs and of external musculature, as well as surgical operations. Compared with Fabrici's record of teaching, however, they sharpen the contrast between Fabrici and his public demonstrations, on the one hand, and Casseri and Galeotto's private ones, on the other. While both Casseri and Galeotto employed a pedagogical approach that treated the procedures of dissection and the internal and external parts, Fabrici used the public demonstration to explore the natural philosophical extensions of anatomy, that is, giving an account of an anatomical part that was single, general, and universal and called for contemplation and an understanding of the organic soul. Much of the instruction that students sought on anatomy, on dissecting techniques, on surgical maneuvers, and the like could not be found in Fabrici's public demonstrations of anatomy.

The students continued to praise Galeotto's style of demonstrating, his manual dexterity, and his responsiveness to his students, as well as the learned content of his teaching. In late January 1586–1587, Galeotto gave a demonstration, amid "a throng of spectators," in the church of St. Catherine, which was appreciated because he had shown a special concern for his students.[108] He received funding from both the Italian and the transalpine students to pay for the construction of the temporary theater without any hesitation on the part of these students. While it was typical for students to pay for lessons, especially private ones, the transalpine students seem especially eager to pay, both because their payment reflects their commitment to anatomy and perhaps because of a perceived favoritism on the part of Galeotto:

> [Galeotto] announced that some Italians had collected out of gratitude some contributions for the construction of the anatomical theater, he indicated clearly that he expected the same from us, especially since he knew that our Nation, beyond the others, delighted most of all in anatomical studies, and since the whole matter would be useful not to him but to his audience.[109]

As this transalpine student recorded, the promises Galeotto made "had been so generous" that the students "awaited nothing the whole winter except anatomical exercises."[110] While the university funded Fabrici's demonstrations in the permanent theater, both the Italian and the transalpine nations were willing to pay for Galeotto's demonstrations. In praising Galeotto, this student noted that the "whole matter" of his demonstration "was useful not to him but to his audience," suggesting by contrast that Fabrici's demonstrations tended to be oriented around his own topical interests rather than the interests of his students.[111] The student explained that "once the start [of the demonstration] had been made, after the external lineaments of the body had been explained, the Anatomist [Galeotto] in the account of the cadaver went straight to the eye, and lectured so clearly and so learnedly on its action and structure that he conducted himself as a man most experienced in anatomical matters and expert [peritissimum] in those of optics."[112] The student's preference for Galeotto should be seen in a context where both Fabrici and Galeotto treated the anatomy of the eye and the subject of optics. Whereas Fabrici's demonstrations foregrounded natural philosophy and the category of use, Galeotto's demonstration began with dissection and with an explanation of structures and functions before moving ahead to philosophical points, that is, to the relationship between the anatomy of the eye and the study of optics. In 1588 and again in 1591, the transalpine students recorded their desire to commend the "generous industry" of Galeotto in any way possible, both in the current year and in all subsequent years.[113]

Fabrici, we may imagine, was not happy. During the following year (1587–1588) he contacted the *riformatori* to remind them of the prohibition against beginning a private anatomy before the public one was finished—suggesting that in the previous year the students had skipped his demonstration in order to attend Galeotto's. When the students learned of Fabrici's underground activity, they described his negligence more fully:

> On the basis of this privilege, which he had obtained to the detriment of the scholars [students] with specious arguments in order to conceal his usual negligence from the *riformatori*, the public anatomist [Fabrici] seized the opportunity and destroyed the anatomical theater of the most excellent teacher, Paolo Galeotto, although it was built at the cost of the students; but without doubt, if we had learned of this more quickly, he would not have dared to undertake what he did not hesitate wantonly to carry out after he had obtained the privilege by trickery [*malis artibus*].[114]

As this passage reveals, Galeotto's previous demonstration, while private, was exceedingly well attended and, because this contretemps took place in the following academic year, Galeotto's temporary theater was still in existence, behaving, in other words, as a permanent theater. In addition, the temporary theater was known as Galeotto's, revealing not only the favor the transalpine students accorded him but also the associative link between spaces and teachers. Fabrici sought to have Galeotto's demonstrations prohibited and *his* theater destroyed, since both presented significant challenges to Fabrici's position and success within the academic community.

The struggle between the transalpine students and Fabrici did not end here. At the end of November 1587–1588, the transalpine students went to the vice-rector of the university to discuss the question of anatomy. Their records explain that they approached the vice-rector about "a place even in private" that could be used for anatomy, though their request was ignored. As to the question of a public anatomy, the vice-rector received from Fabrici "as much in this year as his predecessors had achieved from him in the previous years"; in other words, "he always brought back the singular and same response in which he promised to administer an anatomy in this very year, but failing to provide certain specifications of the time or other circumstances."[115] Fabrici's lack of enthusiasm and lack of commitment were directed at the annual anatomy, not at the other parts of his research or at his writing and his publication project. The exchange—and the circumstances—provide a clear contrast to figures such as Vesalius and Falloppio, who eagerly and repeatedly sought opportunities to dissect.

The transalpine students, once they "caught scent of his deception [*nos vero cum fraudem hanc suboleremus*]," pressed their case further. Eventually, five or six transalpine students, along with the vice-rector, went to Fabrici's chamber, waiting over an hour for him to arrive before they asked him to conduct the anatomy: "We begged first by all the others and then indeed especially by the transalpines (many of whom recognized the benefit of the proficiency of this one alone) that he make a beginning of the anatomy."[116] They also recorded Fabrici's reply:

> After mentioning his great affection for the transalpine students (whom he was not in the habit of deceiving except with flattering words), he gave us hope for the conducting of an anatomy even in this year; truly, we, who had learned from the records of our predecessors not to buy hope for a price, were not content with these words; and we asked about the time and other circumstances, to which he responded first one thing and then something different.[117]

The students expressed doubt about the "celebration of the anatomy," not only because they knew Fabrici was deceptive, using trickery to hide his negligence, but also because they had gone back in their own records and had seen as much. They were evidently committed to the study of anatomy and to their identity as seekers of anatomical knowledge. They were also committed to the documentation of anatomical events, because they subsequently explained that the details of this case could be found in the records of the University, "which have been placed in our treasury [or storeroom]."[118]

This struggle ended with Fabrici giving an anatomy demonstration, but he did not go quietly. Fabrici told the students that he would conduct the anatomy, although, "truly backed into this corner [*in has angustias reductus*], he began to form a little anger" and said "you press so hard for an anatomy while you have provided neither *massarii* nor cadavers."[119] Fabrici then said that he would choose the *massarii*, whom he called *anatomistae*. The transalpine students knew that this was "plainly in opposition to the statutes of the University, namely [chapter] 18 of book 2," but, "lest he move some mob against us, and lest we ruin the fixed time of the anatomy with brawls and struggles," they decided to go along with it.[120] Fabrici promoted "four young men who frequented the Jesuit school," and two were chosen from these four. As the transalpine students explained, however, these candidates were not only entirely inexperienced in "this skill [of dissection]," but also ignorant of all internal medicine.[121] Fabrici conducted the anatomy, but he chose inexperienced hands, a lesson that was meant to admonish the transalpine students for overstepping their station. The lesson also reveals a more important point, namely, that the student-assistants, who were regularly being called *anatomistae*, were supposed to be competent and embody more expertise than their colleagues, suggesting that they not only prepared the cadaver for dissection but completed some of the dissection as well. The students had the last word, it seems, because on 25 January 1588, the transalpine students, along with a student from Verona by the name of Jacopo Donato, went to present their "clear arguments" to the *riformatori* about Fabrici's negligence.[122]

During the same year (1587–1588), the president of the transalpine nation, Petro Paulo Hochstettero, uncharacteristically commented on the medical courses (*concursi*), which were run efficiently and with a decidedly practical emphasis. After Girolamo Capivacci taught the "most exquisite doctrine of the diseases of the head," he turned to the diseases of the eye and "with his great erudition always showed all these things to us to be connected and not obscure."[123] Meanwhile, since he was leaving for Bologna, Girolamo Mercuriale gave a farewell lecture, which the students greeted with "many tears" and "not without sadness"; his absence was seen to curtail the flowering of the university, "the market of all the sciences."[124] Albertino Bottoni lectured on the causes, signs, indications, and prognosis of diseases, "setting forth not only the material of the most effective [*accomodatissimas*] remedies but also some things of mystery [*secretis*] that he had added."[125] The success of his teaching made him "a man of singular good will and humanity"; the transalpine students "followed [him] for many years." Bottoni apparently lectured clearly on all things related to theory, and "in practice" he was "accustomed to invite us once more to [attend to] the marvelous [*mira*] felicity and most accurate method of healing."[126] Marco degli Oddi lectured for several days at the hospital of S. Francesco on the topic of urine analysis and the differences in urine, including those attributed to sexual differences.[127] This list reflects the student's interest in practical courses and texts, in the extension of practical into theoretical matters, and in the secrets, or mysteries, of nature. During this year and in the years to come, the practicality of these lessons was seen to contrast with the impractical nature of Fabrici's public demonstrations. According to a spurious text, *Lamento del Bò*, this was the year that Fabrici privately dissected and demonstrated the eye and the ear, subjects that were to form an important part of his subsequently published research.[128] However, the demonstration was soon suspended because, in beginning the dissection of the vocal organs, Fabrici lost his own voice.[129]

The students' interest in practical courses meant that whenever they found utility in the anatomy demonstrations, they noted it in their records. In 1588–1589, Galeotto gave a private lesson *in officina al Corallo*. This lesson, which was again mentioned in the records for the subsequent year,[130] took place in a pharmacy (*ad Corallam*), a setting of considerable practical importance to the study of medicine:

> When . . . the public dissection [was] finally brought to an end, the most excellent man D. Paolo Galeotto, lest he in any way seem to have retreated from his former desire to deserve well from the students (even though his own theater had been destroyed by the injustices of the Public Anatomist [Fabrici], and he himself had been harmed in various ways by the same man, and had been put off to the less convenient period of Lent), conducted in the workshop *al Corallo* for three whole weeks a thorough and complete anatomy, in which not only did he demonstrate most clearly and with amazing facility and beautiful method the way to dissect bodies, the structure of all the parts and their actions and functions, but he also showed us the ways . . . through the whole body of the veins, arteries, and nerves, to the great delight of us all, and without any expense to us (although he urged that it [funds] should be collected from us as is fair from spectators).[131]

Undeterred by Fabrici's ploys and willing to wait until Lent, Galeotto gave what the student described as both a "thorough" and "complete" demonstration. It covered not only the anatomical structures of the entire body, including the paths "through the whole body" of the veins, arteries, and nerves, but also the method of dissection. While the framework of this demonstration appears similar to the one employed in Fabrici's work—the format moves from structure to action and use—it is crucial to note that the student writing the report repeatedly emphasized structural detail and the comprehensive nature of the demonstration.[132] When Fabrici gave the public demonstration that year, it only minimally satisfied the students. Their lack of zeal was due to the incomplete nature (or topical focus) of his demonstrations: one on the formation of the fetus in utero, or the topic of generation, and another on the muscles of the larynx, or the topic of speech.[133] In the following academic year (1589–1590), Galeotto gave a private anatomy that students described as "briefly succinct" on "all things of complexity": "[he] passed over nothing . . . and he exhibited perfectly the ways of the nerves, arteries, and veins to us."[134] By the end of the 1580s, the students had developed a strong attachment to Galeotto and a growing admiration for his pedagogical style, his treatment of many parts of the body, and the clarity with which he explicated them.[135]

Students continued to seek private instruction, and to associate it with the opportunity to see (and perhaps practice) surgical operations. In December 1590, Vittorio Merullio da Saxoferrato, the vice-rector of the *artisti* students (those studying subjects other than law), wrote to the *riformatori*; in his letter, he acknowledged that alternative lessons in private anatomy were "more useful" than the public demonstrations.[136] Useful, here, did not refer to the Galenic category of *utility*, but instead to an experience more closely tied to the cadaver—instruction on surgery and morphology. Merullio da Saxoferrato may have been comparing Galeotto's demonstration with Fabrici's, for, in the same year, a transalpine student again complained about Fabrici's demonstration:

> [Fabrici] has already spent two months on the exposition and description of the bones of the head. Having turned to the muscles he has completed three, devoting one hour to each muscle. There are so many muscles that, proceeding in this way, two years will not suffice. So when then will he deal with the viscera? In addition, everything is treated confusedly and in a disorderly way: once he discussed the detached arm, before going on after many days to discuss the foot. I don't see how anyone can learn the sequence and connection of the whole from looking at these.[137]

The students were subsequently asked, "Why do you desire a private anatomy, do you not find the public one pleasing?" They responded that Fabrici "provided a most exact anatomy, but we ask that what the anatomist delivered profusely in an informal way of speaking be put before us in the form of a visual synopsis [such as a chart] that can be remembered, as we do in the private anatomy."[138] Fabrici subsequently refined his delivery and formalized his way of speaking. Moreover, not only did students distinguish between teachers and between pedagogical styles, they also understood that the distinction between public and private anatomies included the difference between hearing and seeing (especially charts). While it is often assumed that all anatomical events depended on and trained students in observation, the practices associated with a new visuality were in the process of development, byproducts of the exchange between professors, pedagogies, and classrooms.

While Fabrici may have been jealous of Galeotto and the support he earned and enjoyed from the medical students, Fabrici's negligence extended to his teaching responsibilities. In 1590–1591, the students were promised an extensive, memorable anatomical demonstration: "As the cold months and the opportunities for anatomical administrations drew near, the Anatomist [Fabrici] promised a most exact anatomy and one lasting four months . . . D. Johann Hertelius called Syndic of the University and Marcus Antonius Ponderanus from Crete serve as *massarii*, offer their labor in obtaining cadavers and procuring anatomical necessities, and are encouraged in their initial efforts."[139] When Fabrici failed to commit himself to the event, even when prodded by the *riformatori*, the students turned to Galeotto, who gave a private anatomy (*extraordinariae anatomiae*) on the 22nd of January that was "without confusion."[140] Galeotto, the students wrote, "put forth exactly the names of all the parts" and was seen by all to offer a "most perfect" anatomy.[141] When the students approached Fabrici about the annual public anatomy in 1592–1593, he used the lack of a theater as a reason to refuse a student's request for the annual demonstration. The student writing the report knew of the rumor that the theater would "by no means" be restored and, when he approached Fabrici, Fabrici replied that until the theater was rebuilt, "he wanted nothing else," that is, he would wait to administer the anatomy.[142] Such instances display Fabrici's lack of concern for his teaching, a lack that, according to the students, also informed the content of his demonstrations. Equally, his tactics reveal his aggressive attempts to control anatomy demonstrations—whether public or private, these events were the sites for the demonstration of anatomical knowledge and thus the sites where reputations and authority were established—and help to explain why, when the second permanent anatomy theater was built, it had Fabrici's name attached to it, literally engraved above the entrance.

## CONCLUSION

From the responses of sixteenth-century medical students, we learn something of the variety of pedagogical styles in use in the second half of the sixteenth century. Students praised Galeotto and Casseri for their technical abilities in dissection and their sensitivity to the educational needs of the students. In contrast, students called Fabrici's demonstrations incomplete, inaccurate, confusing, and, eventually, useless. Their responses allow us to chart not only the intellectual coordinates of Fabrici's project—how and where it developed—but also the institutional context that shaped parts of its development.

This institutional context included protocols, expectations, competition, and administrative pressure (applied most forcefully by the *riformatori*). In the 1570s and 1580s, the students complained about Fabrici's teaching, fueling a series of reforms in anatomical education. These reforms made private exercises a more frequent, regular occurrence at the university; they dictated that demonstrations and lectures were separate but necessary to each other; and they conditioned the construction of the first permanent anatomical theater in 1583–1584. With these reforms and with the natural philosophical features of Fabrici's anatomical program, the Paduan tradition of anatomical inquiry and practice became recognizable as something particular and distinguishable from its European counterparts.

The curriculum continued to separate the demonstrations, which revealed anatomical structure, from philosophical lectures, which emphasized explication and commentary on ancient texts as well as the relationship between anatomy and natural philosophy. In 1592, Fabrici dedicated the lecture to his anatomical method before giving a brief private dissection of the eye and another public lecture on vision.[143] Because he focused the lectures on his anatomical method and on the relationship between anatomy and natural philosophy, and because he aligned this discussion with his demonstrations, Fabrici's students were understandably confused. They complained that Fabrici's demonstrations were incomplete, because they focused on the philosophical components of anatomy rather than on the technical or comprehensive aspects of human anatomy. Thus the content of his demonstrations and that of his lectures became indistinguishable.

The first permanent anatomical theater is at the center of this story. This theater was dedicated solely to Fabrici and to the instruction he offered. In Leiden, by contrast, the anatomical theater became a museum in the off-season; its theme was *memento mori*, and it fostered a culture of guides, who made their living by selling guidebooks.[144] In Padua, however, the first permanent anatomical theater was a sign of Fabrici's prestige in the academic setting of Padua's university. It allowed him to compete with Galeotto and with the tradition of private anatomies that was gaining strength (and supportive students). The brief existence of the first anatomical theater—it remained standing for eight years, but was in use for only five—confirmed the link between teachers and demonstrations. When Fabrici brought his complaints about Galeotto to the administration, he made the existence of alternative theaters pivotal to his argument, and he used the lack of a permanent theater as a reason to postpone his own demonstration. In 1594, when the second permanent anatomical theater was almost complete, a transalpine student approached Fabrici with concerns about the annual public demonstration, remembering the "tricks by which he agitated the University" and that his "early skill . . . was now hardened." Fabrici himself mimicked the typical responses of the students, stating, or "spewing forth," "that he could teach or demonstrate nothing brilliantly, nothing usefully, nothing productively."[145] As the students began to emphasize the profitable instruction gained in private settings from other anatomists, they, conversely, came to expect minimally useful public demonstrations from Fabrici. As the next chapter will explain, that shift coincides with a development that took place in the second permanent theater: a dramatic *and* philosophical tradition of public anatomy demonstrations. The second anatomical theater celebrated a more coherent, institutionalized form of anatomy, a spectacle of anatomy that positioned Fabrici as the spokesman for the new philosophical discourse of anatomy and that consolidated and epitomized "the glory of Venice."[146]

# CHAPTER THREE

# Civic and Civil Anatomies

## The Second Anatomical Theater

I n 1595, the second permanent anatomical theater opened in Padua. Similar to the first permanent theater, it was situated in the central building of the university, the Palazzo del Bò, but, unlike the first, the second theater was a larger construction. The space was desperately needed. In 1588–1589, the last year in which the first theater was in use, the students described the theater as dangerously overcrowded.[1] The anatomy demonstration that year relied on two male corpses, the bodies of criminals executed for their "evil deeds."[2] On 23 January 1589, seven days into the demonstration, just as the anatomist Girolamo Fabrici began his afternoon lecture on the topic of generation and the formation of the fetus in utero (a topic obviously not tightly linked to the male corpses on view), the theater was "besieged by many people, friends of the student-assistants [massariorum amicis] and members of the local community [popularibus]."[3] Although the beadle tried to lead the doctors and professors into the theater, many people arrived late, and the noise and confusion interrupted Fabrici's lecture.[4]

Crowd control, however, was not the only issue. A transalpine student reported that the beadle tried to handle the latecomers and especially the syndic (a university-educated man who served as a legal overseer and sometimes as a secretary for the rector;[5] see figure 5). The syndic could not make it to his usual seat near the doctors and professors (a placement decreed by the university's rules) and had tried to stand with the lead counsel of the university; the beadle ordered the *massarii* to give up their seats in order to give the syndic a place. Just then, two of the *massarii* who happened to be Sicilian stopped the show. The transalpine student writing the account said he "might more accurately say, two assassins," but they were students.[6] These Sicilians ignored "the authority of the entire university and the most excellent who were present" by interrupting the event and threatening the spectators.[7] Amid "hisses and disgraceful words [*sibilis et ignominiosis verbis*]," a fight broke out:

> When the anatomist [Fabrici] realized that the matter was coming to arms and that this small dispute was fanning into a conflagration which would be difficult to extinguish later, and also that because of his proximity to them, some harm might also fall on him for this, he immediately and earnestly forbade the Sicilians to undertake anything further against the Syndic.[8]

In the end, the thuggish Sicilian students were removed from the theater, though the barely contained violence of the event helps to explain why this theater was never used again. It was destroyed sometime between 1589 and 1591.[9] Following the conflict, officials decided that at future demonstrations, some students should be better armed so that they would be able to intercept and resist potentially violent adversaries.[10]

The episode illustrates the logistical difficulties and potential excess involved in staging a public anatomy demonstration. A larger theater was certainly needed for the event's growing audience. In light of the way scholars have connected public anatomies to the rituals of Carnival, this scene might be taken to reflect the lawlessness of the Carnival season.[11] To generalize that into a model for Padua's public anatomies, however, would be a mistake, because nothing about this episode was typical of the Paduan tradition. Students, for example, were rarely told to carry more weapons into anatomical exercises or any other lessons. Moreover, while the audiences for public anatomy demonstrations were usually limited to students, professors, and a few magistrates, this particular demonstration appealed to the friends of the student-assistants, and perhaps to surgeons and apothecaries as well as to charlatans and butchers; it also

FIGURE 5. Jacopo Tomasini, *Gymnasium patavinum* (Udine: Nicolai Schriatti, 1654), 57, rector and syndic. Digital image, M0016099. Reproduced courtesy of the Wellcome Library for the History of Medicine, London.

attracted members of the local community. Finally, while the first permanent theater had been the location for disputes between professors and disruptions caused by students, it had never been the site of overt aggression.[12] These transgressions, though, were foremost in the minds of the transalpine students. Once the event was over, they focused again

on questions of comportment and etiquette, asking the syndic to intervene on their behalf with Fabrici and smooth over any perceived offense,
perhaps because Fabrici might have mistaken the transalpine students'
friends as the cause of the disruption and held the nation at fault for the
debacle.[13]

The threat of violence and its repercussions were due, in part, to the
fact that this demonstration was delayed. Annual anatomy demonstrations were supposed to take place during the winter break, after Nativity
but before Carnival. In this year, however, Fabrici's demonstration took
place during Carnival.[14] Carnival was a time of transgression, a time when
hierarchical power structures could be inverted or ignored, when bodies
were exposed and often violated, and when the social order revealed its
fault lines.[15] In a more pedestrian sense, the Carnival season was known
not only for its bawdy escapades but also for an increase in crimes, rang-
ing from petty theft to sexual exploitation and murder.[16] More than simply a small bureaucratic failure, the schedule change allowed this anatomy demonstration, an academic ritual, to overlap with the annual Carnival
and its civic rituals, and this overlap was experienced as a collision. The
theater was packed; academic and local audiences mingled; the crowd-
ing demanded that the seating arrangement be amended: and the Sicilian *massarii* were asked to give up their place to the syndic. The event exposed the failure of the anatomical theater and all those involved in the
staging of the demonstration to control a mixed audience by maintain-
ing the distinctions (and the seating arrangement) that brought stability
to a fragile social order.

While public anatomy demonstrations and anatomical theaters have
repeatedly been linked with Carnival—with its inverted or subverted power
structures and transgressed boundaries of propriety—and with crime
and punishment, this is the only account of such an alignment in Padua.
Its atypical status and the ensuing problems, delays, and frustrations of
administrators and anatomists, however, made it influential. Indeed, the
case served as an incentive not only to reconsider public safety but also to
formalize the annual anatomy demonstration and, thereby, to distance
it from Carnival. It prefigured the development of a highly formal tradition of public anatomy demonstrations. That formality can be detected
in the construction of a new, larger anatomical theater, in its fancier decorations, in updated protocols and procedures, and in the students' concern over their behavior. The event also emphasized the relationship

between anatomy and natural philosophy. As chapter 2 explained, this was a major goal of Fabrici's research, and it further elevated the status of the anatomy demonstration. This chapter explores the inception and composition of the new formality around public anatomy demonstrations; these details hold the key to understanding the dramatic nature of the anatomy demonstration—its spectacular force—and its impact on medical students. The spectacle of post-Vesalian anatomy was not based on bodies being cut open or on the translation of bodies between academic rituals and the civic ones of Carnival. Instead, it borrowed and fused elements of rhetoric, poetry, and stagecraft, offering an intensely aural (rather than visual) experience. Inside the anatomical theater, the spectacles of anatomy commanded attention as philosophical and contemplative events.

This kind of dramatic expression highlighted natural philosophical themes, but it tended to obscure ideas and practices related to dissection. Public anatomy demonstrations, as they developed their formal qualities, largely removed the processes of dissection from the exhibition inside the anatomical theater. This mediated fears about dissection, fears which reflected a continued concern for burial rites and the acquisition of cadavers, not from the gallows, but from graves and local hospitals.[17] While chapter 4 follows that concern as it migrates to the private arenas of anatomical study, this chapter focuses on the new and heightened sense of formality attached to public anatomies. In Padua, the anatomical theater was not the site of raucous, bawdy, or even bodily display; corporeality remained a more prominent feature of the celebratory, often violent escapades of Carnival. Stable matriculation patterns and the fact that the annual anatomy demonstration was usually held in the winter months before Carnival began meant that professors, administrators, and students attached a different set of ideas to the anatomical theater; these were organized around the importance of natural philosophy and the civic recognition that the study of anatomy received.[18]

By hosting more transparent, formal, and dramatic events, the anatomical theater distanced concerns about the illicit acquisition of cadavers. It also activated concerns about comportment and the ways that students should behave, particularly when they studied anatomy.[19] In the episode above, the transalpine students worried that Fabrici would find them at fault for the Sicilians' disrespectful behavior. The syndic's behavior also speaks to issues of comportment, for the record notes that he wisely opposed the Sicilians and "did not suffer himself to be moved from the place [or argumentative position] which he occupied even by the breath of a fingernail [*neque latum quidem unguem*]." We would understand the breadth of a fingernail as a spatial description, but the sources for the phrase—one was Plautian comedy and the other was the Bible—suggest a kind of theatrics of commitment and an unwillingness to fall away from the object of devotion.[20] It is a perfect, if slightly frustrating, combination of ancient and Christian ideology, and it leaves us with a sense of how the theater brought out aspects of comportment even in dysfunctional demonstrations such as this one. Elsewhere, and with a similarly specific reference to the theater, students characterized themselves as "modest" and "civil"; and they adopted this posture when interacting with professors and sometimes with corpses.[21] Understanding these developments allows us to correct the view of public anatomies as gory experiences and to see them, instead, as widely engaging expressions of natural philosophy. Nonetheless, these expressions also had a regulatory, or disciplinary, valence. In the second anatomical theater, the demonstration became a civic and civil event.

THE NEW ANATOMICAL THEATER: *Architecture*

The second anatomical theater was more magnificent than the first in every way. According to an early chronicle, the second theater was built in 1594.[22] Completed and in use by 1595, it was an elliptical arena, a wooden stadium with surprisingly sheer verticality (a feature emphasized in figures 6 and 7).[23] Stairs encircled the shell structure and provided a means of access to the six elliptical tiers (figure 8). Although there were eight windows, which also existed in the previous building at the Palazzo del Bò, these were blocked up until the mid-nineteenth century.[24] The theater was not illuminated by natural light, but rather by torches and candles, an aspect depicted in an eighteenth-century etching of the theater's interior (figure 9).[25] It has been estimated to have held approximately 240 people.[26] The theater celebrated the academic tradition of anatomy and witnessed the emergence of a dramatic tradition of public anatomy demonstrations.

Despite the literal meaning of *theatrum* as a place for seeing, this anatomical theater did not foster the visual apprehension of anatomical particulars. Instead, it provided and cultivated a nonvisual, predominantly

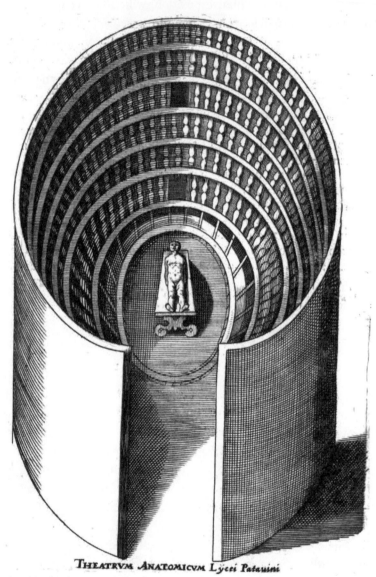

FIGURE 6. Jacopo Tomasini, *Gymnasium patavinum* (Udine: Nicolai Schriatti, 1654), 74, anatomical theater. Reproduced courtesy of the Boston Medical Library in the Francis A. Countway Library of Medicine, Harvard University.

FIGURE 7. Jacopo Tomasini, *Gymnasium patavinum* (Udine: Nicolai Schriatti, 1654), atrium, anatomical theater, and list of lectures. Digital image, M0016104. Reproduced courtesy of the Wellcome Library for the History of Medicine, London.

aural or auditory experience of anatomy. The evidence for such an interpretation is based on the architecture of the theater and on the kind of material that Fabrici presented.

The most important architectural feature of the theater was the one noted by Cesare Malfatti (b. ca. 1550) in 1606. Malfatti described the theater as beautifully constructed in walnut, containing two rooms: one for the anatomists and the demonstrations, and one for the cadavers and their dissection.[27] In his history of the university, *Gymnasium Patavinum* (1654), Jacopo Tomasini (1595–1655) extended this description: the

FIGURE 8. Photo of the interior of the anatomical theater, Padua. Digital image, M0015500. Reproduced courtesy of the Wellcome Library for the History of Medicine, London.

second anatomical theater contained two chambers, an inner one where "the cadavers and the dissected parts were worked upon," which also contained "skeletons, instruments, and other things likely to be used by the anatomist," and an arena where the cadavers were "demonstrated publicly before the presence of various student nations," university officials, and statesmen.[28] The inner chamber is probably the space beneath the theater, a stone basement in which the cadavers could be prepared before they were passed on to the anatomist in the theater; stone would have kept the cadavers cold and thus retarded the processes of decay.[29] The separation of the two chambers, one for preparation and one for exhibition, meant that the demonstration privileged the display of a previously dissected cadaver rather than the process of its dissection. The point is simple, but it has profound consequences.

First, while the display depended on seeing the cadaver, it did not lend itself to scrutinizing the structures and shapes of anatomical particulars. The architecture of the theater, its arena, constrained movement.

TEATRO ANATOMICO DI PADOVA

stabilmente eretto nel 1594.

FIGURE 9. Pietro Tosoni, *Della anatomia degli antichi e della scuola anatomica padovana* (Padua: Tipografia del Seminario, 1844), title page. Reproduced courtesy of the Boston Medical Library in the Francis A. Countway Library of Medicine, Harvard University.

The rows are narrow. As E. Ashworth Underwood suggested many years ago, once a spectator entered the arena, he could not have easily moved past other spectators in the row. Underwood also noted that the spectators in the uppermost gallery would have been twenty-five feet from the cadaver; thus "a considerable proportion of the students must have had to take the professor's word for what he was doing."[30] Similar to much of Renaissance drama, these anatomical performances relied on verbal skill (that of the anatomist) and auditory apprehension (by the audience).[31]

Second, the role of seeing was diminished, because the rows in the theater closest to the cadaver were reserved for the dignitaries in the audience. These included Venetian officials, university administrators, and professors.[32] This left the upper rows and sections of the theater for students, an arrangement that compromised their sight lines and their ability to see many of the dissected parts. While this arrangement is consistent with the university statutes governing all university assemblies, convocations, voting, and the like, it reinforces the idea that the anatomical theater did not foster or enhance the practices associated with visual scrutiny, focus, or observation. Such features required the development of new practices and new (private) spaces.[33]

Third, because of this architectural separation, the demonstration inside the arena did not emphasize the process of dissection, which was done mostly in the antechamber or the basement before the demonstration began. Moreover, in 1595, the year the theater opened, the student-assistants (*massarii*) were renamed *anatomistae* (anatomists), implying that the process of dissection was not only prior to and separate from the demonstration, but also part of their responsibilities.[34] This implication is borne out in subsequent descriptions of their activities. For example, on 20 November 1600, the "anatomists [*massarii*]" were elected and, "for the Excellent Anatomist [Fabrici]," they "were accustomed to offer their hands as helpers in the procuring and dissecting of cadavers."[35] This remarkable feature also suggests that the dissections done as preparation were not intended to be expert ones; they didn't need to be exact.

While the architectural features of the anatomical theater complicate any straightforward account of it as a "place where one sees," they support some of the pedagogical goals and tendencies of Fabrici's lessons. Fabrici expounded the anatomical causes of nature; in his public demonstrations, he neither focused on the particular details of anatomical structure nor on the procedures of dissection. The students recognized this. In 1588–1589, just before the disruptive Sicilian students were asked to give up their seats, the transalpine students described Fabrici's demonstration, noting that he had set about dissecting the structures of the human body: "Although Fabrici might have taken up the administering of such an anatomy for nineteen consecutive days, and sustained himself somewhat by diligence and [the knowledge that came from] his many years," these students thought that "he hardly kept his promises."[36] In other words, though Fabrici might have focused more attention on the dissection of the body and its particular structures, he did not. Nonetheless, this was consistent with his program of anatomical inquiry. As chapter 2 explained, Fabrici's program was based on Aristotle's works and was intended to develop a philosophical anatomical account of the Whole Animal. Unlike his predecessors—Andreas Vesalius, Realdo Colombo, and Gabriele Falloppio—who covered the bones, the muscles, the arteries, and the organs in their demonstrations, Fabrici rarely provided a structural account of the anatomy of the whole cadaver. Instead, he focused on topics, on groups of structures, and on how they appeared in a range of animals. He then provided a single, general, universal account of anatomy, which did not depend solely on the visual apprehension of structural detail.

Inside the second anatomical theater, where Fabrici pursued and presented his research, his focus was often on a single topic and a limited set of structures (rather than a comprehensive treatment of the whole body).[37] In the case of the senses, Fabrici's demonstration would treat the eyes, the ears, and the voice complex, but these parts were small in size, incredibly delicate, and susceptible to decomposition. Inside the anatomical theater, these structures could not have been studied closely, but this is less of a problem than it might initially seem. Fabrici's research did not depend on the discovery or isolation of new anatomical structures. Instead, his demonstrations depended on structural specificity, the kind that dissection would grant, but they pursued normative structures, that is, the regular features of anatomy that were present or absent in the cadavers. Here, Fabrici was following Aristotle. Aristotle's study of nature began with experience, or sense perception, but the experience was of a common rather than a rare or particular sort.[38] While the study of common phenomena (or nature as it regularly appears) would eventually give way to the study of particular, highly circumstantial phenomena, the Paduan tradition of anatomy, as it was embodied in Fabrici's teaching and practiced in the late sixteenth century, was dedicated to normative anatomy.[39] The extraordinary and the monstrous lay beyond the bounds of Fabrici's program and beyond the Aristotelian coordinates of explanation. Fabrici was dedicated to regularity: "even if certain extraordinary animals form exceptions, they do not invalidate the truth of my statements."[40] Thus, even if students, in the course of their preparations, failed to isolate the set of structures under review or accidentally destroyed them, Fabrici could verbally describe them and get on with the business of their functions and uses. The structures were normative; they were, for the most part, already known.

### THE NEW ANATOMICAL THEATER: *Audience*

The anatomy demonstration was not a spectacle of corporeal dismemberment; it was instead a natural philosophical presentation, made engaging and dramatic. Tomasini refers to the doors and walls of the "smoothly joined and elegantly crafted" arena, which were decorated with the insignia of the professors, and to the ways in which the anatomist labored, scrutinizing "the secrets of nature [*naturae arcana*]."[41] The secrets of nature did not refer to the marvels or monsters of nature, but rather to its commonplaces.[42] Tomasini did not elaborate on the details of corpses or the processes that were used to display their internal forms. Rather, he (like Fabrici) stressed the philosophical dimensions of anatomy, dimensions that were intelligible against the longer tradition of the secrets of nature.[43]

These features resonated within the broader community. As did the last demonstration in the first theater, the first demonstration in the new theater generated interest among academic and local communities. According to the students, the second permanent theater opened its doors in 1595 to "Jews [perhaps Jewish scholars], teachers, tailors, shoemakers, sandal-makers, butchers, fishmongers, and, lower than these, porters [perhaps funereal ones] and basket-bearers."[44] The transalpine student who furnished this description noted the presence of almost the "whole citizenry" and listed the specific professions of the members in the audience. In addition, he organized his description according to the status of these professions, where, for example, porters were ranked lower than fishmongers, suggestively offering insight into the hierarchical categories that organized this student's experience as a foreigner in late sixteenth-century

Padua. The description, moreover, implies that the second anatomical theater (unlike the first theater) maintained the social order, though a subsequent record indicates that not all students were so settled, nor were they pleased by the presence of nonacademic spectators, as we will see in a moment.

This was a special period in the history of the University of Padua, one marked by a visibly international student body. The number of foreign students in Padua reached a record high at the end of the sixteenth century, but it was a pinnacle that would not be maintained, due to Counter-Reformation politics and the opening of new universities in alpine lands in the late sixteenth and early seventeenth centuries. At the end of the sixteenth century, foreign students graduating with distinction could still hope to obtain one of the academic posts that this university reserved for foreigners. In the seventeenth century, though, the number of posts available to foreign graduates rapidly decreased.[45] In the 1590s, when the second anatomical theater was constructed, transalpine students constituted perhaps as much as half of the students studying medicine, suggesting that the construction of the new theater and the demonstrations held within it concerned these foreign students to a larger degree than has previously been recognized.[46] Inside the theater, students assumed their places with others in their respective nations.[47] In this way, the theater allowed the hierarchy between student nations to assume material existence—the transalpine students had a large, visible presence inside the theater that would have reinforced the importance of their nation within the academic community.

In 1597, Johann Svenzelio, the conciliator (or president of a student nation) of the transalpine nation at the University of Padua, noted the importance of the transalpine students and made special reference to the anatomical theater. He claimed that "certainly" if foreign students (*ultramontanis*), including the members of his nation, had not shown up in the anatomical theater where the public anatomy demonstration was held, then the theater would frequently have remained empty.[48] The claim is both suspicious—the anatomical theater packed a full house just two years earlier when it opened its doors for the first time—and aggressively self-assured. The politics of the student body were as clear to the students as they were to the upper levels of the university's administration, and even to the Venetian Senate. The Senate called attention to the good behavior of foreign students, the "great valor" of the foreign student body, and the ability for both to enhance the reputation of the university abroad, in *il mondo tutto*.[49] Svenzelio realized that public assemblies, such as the anatomy demonstration within the new theater, could instantiate the power of his nation and enhance its reputation. This was increasingly important because, in the face of Counter-Reform initiatives, the transalpine students sought direct protection from professors as well as Venetian authorities, a protection more readily given when the reputation of foreign students was a good one.[50]

The second anatomical theater also attracted members of the local community. Fabrici, especially, welcomed this change. Most likely motivated by the size and diversity of the crowd, in 1595 Fabrici wrote a letter to the *riformatori*. In it, he asked the Senate to subsidize the annual event by providing additional funding so that entrance into the theater would be free of charge.[51] In making this request, Fabrici sought to reverse a very old tradition governing attendance at anatomy demonstrations and establish a new protocol that reflected the degree to which this first demonstration in the second theater had lured people from well outside the academic community. In order to attend anatomy demonstrations, students were required to pay three *marcelli d'argenti* (a silver coin, minted in the age of Marcello and worth approximately 10 shillings, or *soldi*); they also had to have matriculated either into the university or into a nation of the university at least one year previously and be studying medicine.[52] On 12 September 1596, the Venetian Senate effectively overturned this policy; it accepted financial responsibility for the anatomy demonstration, supplying funds so that the anatomist could pay his assistants and the costs of admission; entrance into the theater would remain free and, if any money was left over, it would return to the state purse and be used for maintenance and conservation of the theater.[53] The Senate's financial backing ensured that the tradition of demonstrating anatomy would continue—it made provisions for materials for the demonstration as well as the yearly rehabilitation of the theater. Moreover, by waiving an entrance fee, it also encouraged the new public orientation of both the theater and the anatomy demonstration.

This public orientation of the theater brought transparency to the study of anatomy. According to university statutes, the procedures for acquiring the corpses for public dissection involved the coordinated efforts of the *podestà*, who oversaw executions and obtained official licenses; the rector, who elected two advanced medical students (*massarii*) to oversee the preparations for the event; and the anatomist.[54] These procedures were meant to regulate as many aspects of the annual event as possible. Inside the theater, however, they became more transparent and were seen as signatures of civic virtue. As the Senate explained in the same decree, because "Anatomy is so necessary to Medicine," its study and its dissections brought dignity to the *studio*, and because the new theater would bear the fruits of the arts of medicine, the theater had been built "in a place most stable and highly honored"; it "will not be disturbed, as is the custom, every year by the bad behavior of the students [*maleficio de'scolari*]," and it will bring "no small amount of dignity to our reputation."[55] The theater would enhance the reputation not only of the university but also of the Republic.

### THE ANATOMICAL THEATER: *Performance*

Similar to the theater building itself, the performance of the demonstration had both academic and civic import. Its symbolic function and dramatic qualities were regularly and simultaneously articulated. The annual demonstration began with a procession that underscored the civic dimensions of this yearly event. Echoing terms that he used when describing the theater's design, Tomasini stated that into "the glittering cradle of the amphitheater" (figure 6) came not only the anatomist with his two assistants, but also the most illustrious prefect of the city of Padua, the rector of the university, many professors and doctors of the college, and students.[56] Engraved in white marble above the entrance was a list of famous proctors and procurators, as well as the name of the presiding anatomist, Fabrici of Aquapendente.[57] The spectacular entrance of individuals moving in procession on the ground was framed by the architectural detail overhead, celebrating as well as instantiating hierarchical order.[58] While such entrances into temporary anatomical theaters and into the first permanent theater were hierarchical and controlled by the rector and the student assistants, it is important to note that within the second permanent theater, the typical protocol was transformed into a clearer demonstration of symbolic meaning, reflected in the theater's architectural details and decorations as well as in the processional entrance of eminent figures.

Moreover, it was not seeing the corpse, but rather hearing the exposition that dominated the audience's experience in the theater.

With a reference to music in the anatomical theater, Tomasini flagged the importance of hearing and hinted at an emerging affinity between the theater as a dramatic space and the anatomy demonstration as a dramatic performance. The event (*concursus*), he wrote, was celebrated "solemnly and prepared festively. The entrance of the theater was crowned with the insignia of the most illustrious rectors of the city and the university and anatomists; and occasionally music was permitted for the orations of the doctors."[59] While Tomasini seems to draw the logic of his description from civic processions, which were also hierarchically ordered and often accompanied by music, his reference to the doctors' orations suggests that the role of disputation had diminished. Earlier anatomists, such Realdo Colombo and Gabriele Falloppio, recount particular disputations that took place during their demonstrations; Fabrici, in the early part of his career, does so as well. Here, inside the second theater, disputation was curtailed. In its place a less aggressive, more refined tradition of orations and presentations was developed, and, for the members in the audience, quiet reserve was the norm.

Tomasini took his reference to music from the records of the transalpine nation. As the transalpine students described it, on 12 December 1597, a group of lute players entered the anatomical theater:

> To please the anatomy spectators and to raise them from their sad look, lute players led by the anatomy students had been brought into the theater ([a practice] interrupted in the previous years) ... those musicians were present as well for many days following, and the expenses were hardly to be regretted by those on whom they were imposed ... thanks to this tranquility the anatomical theater will be able to persist for quite a long time unharmed for some years.[60]

While musicians were known to accompany academic exercises—the fame of Francesco Portenari, a musician who also taught music lessons to students, testifies to the interest that students took in musical education—they also served to make typical academic exercises special.[61] With musical accompaniment, the annual anatomy demonstration stood in more explicit relation to other academic exercises: inaugural ceremonies and important addresses, which took place in theaters, as well as other lectures and events attended by academic officials and Venetian grandees.

The presence of musicians helped to generate institutional significance for the event, transforming it from an academic ritual into a special kind of performance. The students, Tomasini explained, were "most delighted" by the music, indicating that this performance incorporated aesthetic pleasure as well.[62]

Music was also functional. It created a tranquil and formal atmosphere inside the theater, creating a contrast with the earlier atmosphere of disruptions and disputations.[63] More speculatively, music may have offered an alternative temporal structure to the event. No longer dependent on the progressive format of dissection, Fabrici could focus on and isolate the areas that interested him. Although the students found the order of his demonstrations disjunctive rather than coherent and comprehensive, the music may have formalized the order, marking the segments of the exhibition as well as augmenting the role of exhibition that the theater's architecture already established. Music could also discourage interruptions, whether they were planned or not. In 1600, the transalpine students again recorded the presence of musicians, as well as a "throng of listeners," at Fabrici's demonstration involving two male cadavers, one female corpse, and "the bones of birds."[64] The students, commenting on the actual demonstration, focused on its ceremonial qualities and the presence of musicians. While keen to watch Fabrici, they also referred to themselves as listeners.

To the extent to which the theater celebrated Fabrici and the institution that supported him, it also accelerated the reputation of the university as more prestigious than other centers of learning, both within the Italian peninsula and beyond it. Historians of the university have noted that the relationship between the university and the city-state of Venice was an increasingly close one, and anatomical practices were a part of that trend.[65] In the early seventeenth century, Tomasini noted that the anatomical theater, along with the public and private festivities associated with civic life, reflected "the majesty and glory of Venice."[66] For Tomasini, the theater and the events it held celebrated, and thus strengthened, the reputation of Venice at home and abroad.

## PHILOSOPHERS AND POETS

Inside the anatomical theater, the performance included processions and music. It may also have underscored the association between the natural philosopher and the poet. This appeared explicitly in the publications of both Fabrici and Cesare Cremonini (ca. 1550–1631) in the late sixteenth and early seventeenth centuries. The connections between natural philosophy and anatomy did not derive solely from the reception of Aristotle's works on nature and on animals. The humanist enterprise (i.e., the intellectual traditions of humanism), as it was developed in the Renaissance, was an educational one that included Aristotle's works on poetry and rhetoric, suggesting an additional Aristotelian influence on the study of anatomy.[67]

The reception of Aristotle's *Poetics* produced various debates about the status and utility of nonmimetic and mimetic discourses, a sort of confusion from which new approaches and inquiries emerged.[68] In the *Poetics*, Aristotle bases his theory of tragedy on the concept of mimesis (the imitation of action) and on the presence of an internally coherent plot.[69] His *Rhetoric* discusses nonmimetic discourses, that is, oratorical as well as dialectical works (which include disputations and expository, scientific, philosophical, argumentative, and historical prose). These forms of argumentation require statement and proof, and they would seem to apply easily to the dialectical traditions of early studies of nature and directly to Fabrici's work and his publications.[70] Yet the dramatic potential of mimetic, poetic discourses lingered.

Toward the end of the sixteenth century, as authors struggled to find unity in the curriculum of rhetorical education and criticized methods for teaching the rhetorical arts, they often elided the distinction between nonmimetic and mimetic discourses. They blurred the boundary between logical disputations and dramatic narratives, between the subject of Aristotle's *Rhetoric* and that of his *Poetics*. In 1574, in his treatise on the orator, Giason di Nores (ca. 1530–1590) recapitulated many of the principles of Cicero's *De oratore*, underscoring the mimetic qualities of good orations, which reveal "a certain observation and imitation of nature"; in his 1586 commentary on the *Poetics*, he began by considering the *Poetics*, like the *Rhetoric*, to be "a subject of civil and moral philosophy."[71]

A similar blend was present in works that established links between poetry and natural philosophy.[72] Bartolomeo Meduna's dialogue on the ideal student, *Lo scolare* (1588), defined natural philosophy as an endeavor to imitate "the great mother of Nature."[73] In the early seventeenth century, Cremonini collapsed the distinction between the poet and the philosopher. In *Il nascimento di venetia* (1617), he explained that "Aristotle has written in the *Poetics* that Poetry comes from Philosophy, for both Euripides and Sophocles were philosophers . . . [and] both Plato and Aristotle affirm that the poet is a force of genius . . . a good philosopher is thus taken [to be] a good poet."[74] In other words, the parallels between poet and natural philosopher were sometimes strong enough to override Aristotle's distinction between mimetic and nonmimetic discourses.

As did Cremonini, Fabrici connected his writing to this broader context, linking his work to poetry and his role as anatomist / natural philosopher to that of a poet. Introducing his work on generation, Fabrici referred directly to the poetical nature of his enterprise. He promised to treat "the first beginnings of human life . . . not the origins of man alone, but those of many other living creatures"; he then asked, "Could one tell or invent a tale more magnificent, more mysterious, or more wonderful than this?"[75] Though this was a rhetorical aside, the question calls upon the figure of the poet and the act of invention, aligning both with the anatomist. Not only does Fabrici compare his natural philosophical study of anatomy to a dramatic narrative, but he also suggests that his study more closely imitates nature. The associations between drama, poetry, and anatomy helped to ground the anatomy demonstrations and the new anatomical theater in the humanist culture of the university, inflecting both with poetic potential.

The emphasis on imitation recalls a passage in Aristotle's *Poetics* (IX.1–4) that explains the differences between history and poetry in terms of imitation. In his 1559 commentary on the *Poetics*, Dionigi Atanagi (ca. 1510–1573) explains the differences succinctly. History does not use imitation, poetry does; history treats the particular (things as they are), poetry, the universal (the pure idea of things); history presents variety (in characters and actions), poetry maintains constancy; history is limited by the material (the truth of the facts), Poetry is not (for it adds things in order to produce the marvelous, the delightful).[76] Fabrici was very much a part of this world of late humanism and its productive consideration of Aristotle's works, so he looked to the similarities between poetry and natural philosophy, focusing on imitation and incorporating a sense of aesthetic pleasure in his work. In deciphering anatomical structures (i.e., particulars, or things as they are), Fabrici extended his account of anatomy and the processes of the soul (such as locomotion, respiration, and generation) into the realm of universals, which are formally similar to the idea of things; these universals maintained constancy across different species and, when presented in the anatomical theater, produced delight across different audiences.

Beginning in 1600, Fabrici published a series of brief tracts, or monographs, on specific processes that, as he said, could be collected and assembled into one large volume. He gave the name *The Theater of the Whole Animal Fabric* to that volume. This "theater" promoted the study of anatomy as a natural philosophical enterprise, rather than designating anatomy as a manual, practical endeavor (useful primarily for surgery). By the early seventeenth century, the two domains of anatomical instruction were recognizably distinct. The curriculum had separated them in 1584, around the time the first anatomical theater was built; by 1595, when the second theater opened, they were instead differentiated by the teachers. Fabrici, with his philosophical anatomies, dominated the public arena; Casseri and Galeotto, with their structural focus and interest in surgery, the private ones.[77] While the intriguing colored illustrations in Fabrici's *theatrum* could serve a number of functions—for example, supplying structural detail that was not covered or not visible in the public demonstration—Fabrici's use of *theatrum* alluded to the theoretical principles of anatomy (as an appeal to natural philosophy), revealing a close relationship between his publications, the permanent anatomical theater, and his demonstrations within it, a concatenation that signals the dynamic interplay among drama, rhetoric, poetry, and anatomy.

Travel writers from this period mentioned the new anatomical theater and highlighted the various dimensions of performance, as discussed above. In his guide to Padua (1623), Angelo Portenari (d. 1624) attributed an ambivalent status to the anatomical theater by describing it as a didactic venue in his section on the Palazzo del Bò and as a dramatic one in his list of the modern theaters that existed in Padua.[78] In his travel guide, Francesco Schott (1548–1622) called the anatomical theater one of the marvels (*maravigliosa*) of Padua: it was a celebrated space used by professors of medicine and an ingredient that enabled the university to be "a market of the sciences," one that rivaled the ancient Athenian academy.[79] In the 1640s, John Eveyln (1620–1706) went to the anatomical theater and noted that the demonstration was "celebrated with extraordinary apparatus [or preparations]."[80] Such accounts attest to the widespread interest in the venue of the theater; they also reveal its dramatic connotations as a theater *qua* theater, as a marvel, and as a site of preparations (or stagecraft). Celebrating the institution of the university and the

Republic, the theater did more than compliment Fabrici's philosophical orientation. It allowed his approach its most dramatic expression, tying particulars to universals and persuading academic and nonacademic spectators to contemplate the secrets of nature.

## COMPORTMENT

For medical students, the anatomical theater and the newly formalized anatomy demonstration contrasted their educational experiences with private dissection. While the students criticized the public forum (an aspect developed more fully in chapter 5), they also sensed how influential this formal tradition of anatomy was. The study of anatomy was recognized as important both by the academic community and by the wider civic one, and it began to generate a range of ideas about manners, civility, and reputation. If the transalpine students point to a general pattern, then students used the popularity of the public demonstration to draw attention to the honor and fame of their respective nations. They also understood that the public venue demanded a certain comportment, guiding their behavior in public, among their peers, and with their professor and the anatomized corpse.

When organizing the demonstration and preparing the cadavers, the student-assistants reflected on issues of comportment. On 3 December 1597, a student described the scene of preparation:

> Although at first they seem to be passionate to begin, suddenly when the navel has scarcely been opened and only parts of the abdomen are visible, they immediately begin to cool off and withdraw, except for a few, for whom modesty and fear of losing the professor's goodwill compel to stay unwillingly to the very end.[81]

This preparatory dissection was done by Fabrici and his student-assistants, who were required to stay by his side. The criteria for their election to the position of assistant are not well specified. The statutes indicate that they must be competent (*duos scholares idoneos*) and have studied medicine for at least two years, but this passage suggests that they had to master (or become master of) their tendency to "cool off and withdraw" from scenes of dissection.[82] What was motivating them to remain, of course, was their desire to please their teacher. Not only that, but the description draws attention to the modesty that compelled the students to stay at the side of the cadaver and, in a controlled manner, continue the dissection.

The Renaissance university has often been seen as a place of aggressive dispute and adolescent pranks and, less often, as a place of pederastic engagements and homosocial bonds. According to the transalpine records, the university was also a place of intense modesty. In 1597–1598, a transalpine student commented on the desire to appear amiable and open, not uncivil (*inciviliter*).[83] While this feature may be more pronounced in the records of foreign transalpine students rather than those of native ones, in the same year we find students voicing praise for Ercole Saxonia (1551–1607). Saxonia was an accomplished student and later professor of medicine and, according to the praise of his colleagues, he was noted for his "propriety [*honestas*]" and "modesty [*pudor*]."[84] In order to understand how modesty might have provided specific constraints in scenes of anatomy, we can narrow the frame of reference to the genre of student comportment manuals. Linking terms such as civility, propriety, decorum, and modesty, these manuals elaborate on modesty from its origins in the rhetorical concept of decorum—where ideas, actions, and speech are determined by the occasion—to its role in systems of manners, where it emerges in internalized notions of discipline. These manuals provided students with shorthand accounts of each of the disciplines as well as instructions for how to organize their daily schedules, how to talk to professors, and how to act when attending university ceremonies. While the manuals were available to most students, foreign students made it a point to obtain them. The transalpine library in Padua, begun in 1586, acquired several vernacular works, including multiple copies of Stefano Guazzo's (1530–1593) *On Civil Conversation* and Giovanni della Casa's (1503–1556) *Il Galateo*.[85] These manuals contain idealized portraits of the well-behaved student and some of the essential rules for the earnest student pursuing his education on foreign soil.[86]

As Biagio Brugi has explained, transalpine students in Padua were notoriously earnest. In 1566, Pope Pius IV issued a bull stating that all university students had to profess their faith in order to receive their degrees, meaning that the transalpine students, the English, and the Greeks, among others, had to profess a Catholic faith. Although the students sought protection obtained from the *riformatori* as well as the doge himself, they continued to be the objects of scrutiny; priests, with the support of local professors, held inquisitions and constrained the students to profess the Catholic faith and to live *cattolicamente*. The level of discipline enacted by these quasi-official persons, along with the fear palpable in the transcriptions of the transalpine nation, serve to demonstrate both the impact of the Counter-Reformation in the Veneto, traditionally known for its official and unofficial distance from Rome, and a partial cause for the transalpine students' concern about behavior at the university.

The comportment manuals, directed at students and consulted by them, adhere to the psychology of youth described in Aristotle's *Rhetoric*. Aristotle writes that young men are impulsive, prone to aggressive behavior and to battle:

> Youth long for superiority, seek it in victory; they are not cynical, because they have not yet learned wickedness; they are trusting, because they have not yet been deceived; they are filled with good hopes, because they have not experienced failure . . . they live for the most part in hope, for hope is of the future, memory is of what has gone before; for young men the future is long, the past short; for in the dawn of life nothing can be remembered, and everything [can be] hoped for.[87]

These manuals, concerned with the psychological development of youth, the dangers of violence, and the ideal of honor, vary in their approaches to the passions—whether to suppress or to rechannel them—and in their commitment to humanist ideals.[88] They unite, however, in the belief that modest behavior, manners, and speech were evidence of a student's nobility. Guazzo, for example, uses *civiltà* to reflect *cortesia* and *politezza*, which contrast with the "customs of men engaged in trade-oriented and mechanical work [*costumi di uomini volgari e meccanici*]."[89] These were evidence of nobility, the very qualities that educators, administrators, and institutions sought to produce, and the keys to a coming-of-age narrative for the early modern period.[90]

The production of the ideal student took shape inside and outside the classroom; it was a process that highlighted old and new sources of knowledge and problems of knowledge acquisition. As the title of Guazzo's work, *On Civil Conversation*, indicates, civil behavior was displayed by civil speech. Guazzo maintained that conversation was better than book learning because conversation addressed the "things about the world."[91] He launched his critique of book learning with medieval scholasticism in mind. Like other humanist pedagogues before and after him, Guazzo sought to provide an ethical foundation for education that derived from classical models; these were supposed to shape a student's moral behavior in the world. However, as he says, "it would be an error to believe that doctrines can be acquired only from books rather than from conversation . . . it is better to apprehend a doctrine through the ears than the eyes, one should not give in to being consumed by sight," but rather (as Cicero indicated in De memoria) "to receive from the ears the living voice, which imprints the mind with marvelous force [mirabil forza]." Guazzo concluded by providing an intriguingly sacramental character to the scene, noting that the soul becomes "languid" and "thin" when it fails to practice the art of disputation.[92]

Offered as a counter to medieval scholastic rehearsals of logic, humanist disputation was meant to reveal, among other things, rhetorical sophistication.[93] During a disputation, two or more disputants would argue for and against propositions in order to arrive at the truth and convince the audience.[94] It was an aural and verbal tradition, and it was the central organizing image of learning in student comportment manuals. Meduna's dialogue, Lo scolare, was directed at the reformation of student comportment but also praised the "spectacles of disputes":

> It is not certain that more is imprinted in the mind if one has heard or read something than in talking and disputation . . . Who then is there that does not see that in order to understand and to penetrate minutely the subtleties and witticisms [argutie], nothing is more worthy than the dispute, used by beloved philosophers, [and] nothing is more enjoyable than the acquisition of the understanding of truth, which one continues to question because, like gymnastics or wrestling, [just as] the body's strength becomes firmer and established, the same [is true] in the battle of letters, [when] the virtues of the soul are rendered stronger and more vigorous . . . the dispute vivifies the mind, makes the memory tenacious, the tongue ready, the soul ardent, and clarifies the truth.[95]

Here, the interlocutor echoed Guazzo and the Ciceronian idea of the voice that imprints the mind; the impression would be retained by the memory and then called forth in the rehearsal of knowledge. Meduna praised disputation while simultaneously noting it as an instance in which one displayed vigor, tenacity, and strength. It was comparable to wrestling, to gymnastics, and, as a "battle of letters," to physical combat. As the descriptions imply, disputation could also produce scenes of boyish excess.

The topic of disputation opened itself to ideas about what constituted civil or uncivil conduct in the academic world. Indeed, following this discussion in Meduna's dialogue, another interlocutor turned the conversation to the topic of friendship, implying that the bonds of friendship might reestablish an idea of civil conduct following the aggressive battles and spectacles of public disputes.

While disputation was a staple of Renaissance education, it seems to have waned in the anatomical theater. The formality of this event instead took a different turn. Concerned about disputation as a site of aggression, other writers on student comportment suggested that within the space where the disputation was held, students should not consider friendship so much as status. Della Casa argued that in a dispute, a student should not be too contrary or too aggressive:

> It is proper to let everyone have his say and, whether the opponent is right or wrong, to abide by the opinion of the majority or of the more importunate and leave the field of battle to them, so that others and not you will be the ones to do battle, work hard, and sweat. These are unseemly occupations not suited to well-behaved men.[96]

While Aristotle understood that young men were prone to aggressive behavior because victory was more important than honor—victory signaled superiority—della Casa cautioned against verbal as well as physical aggression. Where Aristotle suggested the rechanneling of aggression, della Casa called for its suppression.

Worried that disputes might dissolve into displays of aggression, even though more or less controlled, writers emphasized protocols, rules, and other ceremonial features of the disputation. Della Casa suggested that students should be attentive to the status of their interlocutor, reflecting that status in the formal nature of their speech and in the number of compliments they offered.[97] Because Venetians, della Casa explained, are prone to lavish compliments, students should be ready to flatter them excessively. He also (and more frankly) noted that "this habit [of upholding and carrying out ceremony] then, so beautiful and becoming on the outside, is inside totally empty, and consists of appearances without substance and in words without meaning. This does not allow us, however, to change it. On the contrary, we are obliged to abide by it because it is a fault of the times, not of ourselves. Ceremony, however, must be carried out with moderation."[98] While della Casa warned that being preoccupied with formality would limit a student's ability to learn "weightier matters," he reiterated the importance of following formal protocol, even in intellectual settings such as disputation. This was one way of engineering acquiescence in the student.

Disputations threatened the sociability of academic life and the production of knowledge.[99] In an attempt to resolve some of the problems that surfaced in academic disputations, but also circulated more widely in the university—namely, problems of aggression and violence—writers also began to entertain and elaborate on the ideal of the silent student. While the silent student can strike several postures—that of a brooding rebel, that of a virtual sleeper, that of a careful listener—these writers were careful to specify the student's subjection. Meduna's interlocutor described the essential features of the ideal student in precisely these terms:

> The subjection of the student has three components: attention, docility, and benevolence, [he should] be attentive with exercises, docile with his intelligence, and benevolent with his soul, attentive to what is heard, docile to what is meant, and benevolent to what is retained . . . [the student, when listening to his teacher] should not wander in his thoughts . . . [but rather] close his mouth and listen with wonder [maraviglia]."[100]

Grounded in concerns about the aggressive nature of disputations, the traits of silence, acquiescence, and docility became key features of the well-behaved student.

Silence was a sign of submission to authority; speech, a sign that authority had been appropriated. Moments in which students were supposed to speak were carefully controlled, as Meduna's elaborate scheme indicates: "the student must speak when the occasion invites it and spurs it on, when it is not useful to be silent." Even then, speech should be "prudent," "truly and modestly and not crazily and arrogantly" put forth; reasoning should be "grave, clearly in the service of gravitas, and pleasing."[101] If speech does not have these qualities or does not serve the principle of gravitas, then Meduna calls it (and the student) shameful, a disgrace. Elaborating on this point, Meduna explains that the student must neither interrupt the speech of others nor dispute them with "harsh words" or a lack of grace, "nor should he attempt to be superior with his arrogance or put forth falsehoods because it is better to be beaten while speaking the truth than to speak lies in order to conquer others."[102] As questions of civil behavior mingled with those of civil speech, and as the impetus to integrate humanist practices of civil conversation into the academic setting became ever stronger, speech was carefully controlled.

For students, questions of speech and language pervaded their academic experience, and these were notably salient around displays of authority (theirs and their professors) inside the anatomical theater. While these connections were magnified in the second theater—where it was clear that Fabrici would speak and that the students would remain silent—such practices were also present in the first one. In 1589, following the interruption by the two Sicilian *massarii*, Fabrici digressed from the main line of his demonstration.[103] He began with a dissection of the "muscles related to speech" and the organs "serving pronunciation," but he paused to connect the subject to his particular audience. The transalpine students described the moment:

> He began on this occasion of his sermon to ridicule the pronunciation [of Latin] of various [student] nations. In the midst he arrived at our praiseworthy nation, saying, "the pronunciation of the transalpines is hard and slow, since indeed they wish to pronounce [words] while they compress the mouth excessively—this would be the cause—so that always they pronounce awkwardly the *f* for the *v*," and for the record, he began to demonstrate the words: qui ponum *f*inum *p*ipit, *t*iu *f*i*f*it.[104]

Fabrici meant *qui bonum vinum bibit diu vivit* (he who drinks wine lives a long time). The students found his joke—that they slurred their speech and spoke like "drunkards [*insignis vini potatores*]"—humiliating.[105] By their account, Fabrici put forth these words "ad nauseam," and "he felt great joy from these words so much repeated that he was unable to contain his own laughter."[106] Although Fabrici brought himself and the scene to order, the transalpine students felt "exposed" "in the presence of all the other nations."[107] With his joke, Fabrici called attention to the transalpine students as foreigners, flagging their status as linguistic, demographic, and (because their Protestant leanings were assumed) religious outsiders.

The scene was not a grammar lesson at all; it captures the less obvious, social (or socializing) functions of the anatomical theater: its ability to configure the status of various student nations (ascendant or not); its ability to recognize the social relations of the students and their professor; and its ability to connect anatomical themes quite directly to issues of comportment (in this case, a manner of speaking). While speech and language were the overt themes of this lesson, it is equally important to consider that the transalpine students were silent. They did not speak to Fabrici, either to engage him or to dispute his remarks. Being silent was a way of recognizing the power and authority of the professor. It derived from a posture of restraint. In comportment literature, silence is a metonym for physical and mental restraint, and an emerging form of internal discipline. In his *Aphorisms* of 1603, Orazio Lombardelli (1545–1608) described formal education in precisely this way, as a "restraining of the passions [*frena le passioni*], tempering anger [and] moderating one's thoughts."[108] In his student manual of 1604, Cesare Crispolti (1563–1608) echoed these sentiments, recommending that a student, with a book in his lap, try hard to reduce textual ambiguities with the maximum courtesy [*gentile*], that is, "with modesty, without passion and [self-]interests."[109] This was not just an approach to a book or a request for critical reading skills. It illustrated what Crispolti called "a good disposition," a combination of innate gifts and proper training that revolved around docility. For Crispolti, docility was a gift from nature, consisting of "intelligence [*ingegno*], judgment, and memory" that depended "on the organs and one's complexion, which derive from one's parents, from nurses, from the air, and from higher causes."[110] Emphasizing the intellect and education, however, Crispolti saw the intellect's "acuteness of apprehension" as the result of a good, docile disposition: "with much diligence, animals of a fiery nature become meek [*mansueti*] and manageable [*trattabili*]. Thus, it happens that with much study and industry, unrefined and harsh intellects [*ingegni rozi e aspri*] soften themselves and render themselves fit for knowledge."[111] Here was a mind and a style of comportment fully aligned with the climate of university education and suggestively vulnerable to external stimuli. It evoked Guazzo's notion of courtesy, best demonstrated in civil conversation, but it supplemented that with forms of restraint. The silent student, who was a portrait of its embodiment, emerged as its cleanest, clearest representation.

The posture of silent looking requires control of the voice. In this period, control and restraint of one's voice became an important substitute for control and restraint of one's body and one's thoughts.[112] In educational treatises, humanists praised physical restraint, mental discipline, dispassion, and silence. Students exhibiting these features were success stories, exemplars of the docile, acquiescent student that administrators, professors, and perhaps even parents sought to encounter. The formulation of the ideal student resonates with the particular skills of both anatomical and medical assistants. Fabrici capitalized on the opposition between speech and silence in the anatomical theater, using it to enhance his authority, guide the reception on his lectures, and effectively dominate that public venue. His student-assistants were silent—modest in their preparations and quiet in the theater. Similarly, silence was a feature celebrated in medical assistants. For example, in the "Cautele dei medici" of Alberto de'Zanchariis, a medieval tract, the author explains that when the physician (medico) attends the patient, "it is best to stay silent and to make [oneself] the ear [of] the merchant, according to the proverb; in fact it is better to be quiet than to err as Galen says in the proem [of the second chapter] of Metacriseos."[113] Silence serves first as a way to improve diagnosis—the physician needs to listen to the patient—and second, as a guard against error. While the tract included a third instance of silence—"refrain from discussion also of the compensation that the physician must receive from the sick because he must solicit [payment] so that he would be paid according to his directive"—it emphasized silence as a mark of verbal and cognitive restraint. This kind of silence, crucial to the learned practitioner, included aspects of submission and authority: the physician submitted himself to the patient and the patient's account of his own condition, but the physician also adapted, reformulated, and demonstrated his mastery over the patient's statements as he provided a flexible diagnosis of the condition (and demanded payment).

As physicians-in-training, medical students needed to learn to be silent, that is, to exhibit forms of both verbal and cognitive restraint. At the university, they would learn this by playing the role of assistant to the physician or to the professor conducting clinical rounds and by attending the anatomy demonstration and playing the role of assistant to the anatomist. In his treatise on the care of the aged, the anatomist Gabriele Zerbi (1445–1505) insisted on his assistant's silence. Assistants were supposed to imitate the gerontocomos, but "let them be close-mouthed, not talkative or loose in talk, and always alert to spring to their duty, completely free of idleness so that absolutely nothing shall be lacking to the task of extending life [of the aged according to the principle of restoration]."[114] In other words, assistants needed to be strong and agile, in addition to being polite, motivated to care, and capable of controlled speech and silence.[115] For Zerbi, as well as for later writers of the period, controlled speech was nearly synonymous with self-control.

As a feature of medical education, silence was useful in scenes of diagnosis and treatment and in activities associated with anatomical inquiry. Students were quiet in Fabrici's lectures and in the face of the corpses they aimed to dissect. Inside the theater, commotion and interruption were still possible, but the event was more fully dedicated to the audible—to Fabrici's lectures and to the musicians he hired—and not to the visible. Questions of comportment continued to appear in the context of this new anatomical theater and to guide (even to dictate) questions about the apprehension of anatomical knowledge. In 1595, as we saw, Fabrici wrote to the *riformatori* to request extra funding so that the theater could be open to everyone, with the demonstration being free of charge. No doubt he was motivated by the large and diverse crowd that attended the premier demonstration in the new theater.[116] Fabrici's publicity stunt was not appreciated by all medical students, however. In the records of the transalpine nation, the conciliator—Georgio Rumbaum Uratislaviensi (a Silesian)—wrote passionately against the idea of a "free anatomy [*liberam anatomiam*]": "I counsel that there should be no credence given to the opinion of those who try to introduce anatomy for free in this academic institution, a sort of disease of the humanities, and the most open window of sedition and murder, contrary to our ancient traditions that have been preserved all the way to the present age."[117] From this tarnished ideal of nobility, the author then turned to the way that craftsmen in the audience "gaped" at the anatomy: it would be better to keep the theater closed to the wider community than "to allow a crowd of the worst sort of gaping craftsmen and vulgar men to occupy the lower benches—nay! the whole theater—to trample, diminish, and impede that sweetest fruit of anatomy."[118] The transalpine student experienced Fabrici's demonstration as contaminated, because of the presence of these craftsmen. The craftsmen came from lower trades; they were not local, respected practitioners (such as Nicolò Bucella or Paolo Galeotto) who were welcomed into the theater and into the inner circles of the transalpine nation.

This description develops themes of comportment by indicating that the presence of craftsmen inhibited the students' ability to see the anatomy. Students had become used to listening to Fabrici during the public anatomy demonstration; the anatomical theater—newly built, splendid, and extravagant—was an auditory space. This time, however, these students were intensely aware of the public nature of the theater, including whom they were standing near and whom they could see. Inside the theater, noble and nonnoble orders mixed; academic and nonacademic people brushed elbows and maybe even jostled for position. In this setting, the students articulated their claim to the visual apprehension of anatomical knowledge . For students, the presence of these craftsmen was unwanted, but the transalpine students noted the impropriety in terms of comportment: these individuals failed to conduct themselves in the right way, for they "gaped" at the spectacle, preventing the students from "properly" seeing, savoring, or contemplating its "fruits."

This episode should be seen in light of both the public and the private traditions of staging and studying anatomy. Because it took place in a new theater and carried both symbolic and civic weight, Fabrici's demonstration made students intensely aware of the nobility of their education. How diminished would their degree be if the university was too often "open" to everyone? The anatomical theater, by intersecting with the ideals of humanist education and the goals of the Renaissance university, activated this set of concerns. Packed into the claim the students made for visual scrutiny, however, was a trace of the conditions present at a private anatomy. Where did students learn to gape in the right way? They learned in private. The disruptive, mixed, contaminating nature of the public demonstration pushed them toward exclusive anatomical experiences, private events that not only fostered habits of observation (as chapter 5 reveals) but also maintained the nobility of their education.

## CONCLUSION

As its dramatic history indicates, the second anatomical theater ushered in a natural philosophical tradition of public demonstrations that cultivated the growing cultural significance of anatomy. The theater brought drama, poetry, and philosophy to the public anatomy events. The cultural significance of public anatomy demonstrations did not depend on the gradual opening of several cadavers or on the processes of dissection. It was not the visual apprehension of anatomical knowledge that structured these public demonstrations. Nor was it bodily gore or a carnivalesque celebration of corporeality that drew spectators to the new theater. While

these features have been the subject of other studies of anatomical theaters, they are almost entirely absent from the Paduan anatomical tradition. In Padua, spectators were drawn to the study of anatomy by Fabrici's
reputation, his tendency to highlight philosophical themes, and the novelty of the theater itself.

Before the second anatomical theater was in use, the students complained about Fabrici's demonstrations. He had frequently dodged his
responsibilities. Even when he did demonstrate, students called his
lessons useless and referred to his skill as hardened. When he demonstrated inside the second theater, however, students began to accept him
and his style of teaching. In 1596–1597, Fabrici received money from the
spectators in the audience, and the students said that, in this year, he
transcended all other anatomists.[119] In 1597, following the public outrage
over medical students who committed acts of violence and sacrilege in
the anatomical theater, Fabrici offered what the students called a "delightful and illuminating" demonstration on a male cadaver and a preg-
nant female one.[120] The assent of his neophyte audience depended as
much on the nature of Fabrici's research as it did on his rhetoric, his
props, and the stage itself. Inside the new theater, each of these was newly
configured for an audience of students and professors, as well as shoemakers, tailors, and merchants, people who had traveled to Padua to
attend the demonstration and see the new theater. There, Fabrici transformed the responses of his spectators, emphasizing the transition from
observation to speculation, that is, from the actual animal to the principles underwriting the Whole Animal.

The anatomical theater did more than just stage Fabrici's commitment to natural philosophy. In the theater, his students learned about
anatomy and natural philosophy; they also learned how to behave with
respect to their peers, their superiors, and the anatomized dead. The
theater celebrated the ideals of silence, obedience, and the acceptance of
authority; when the students embodied these ideals, the atmosphere inside the theater was one of tranquility. In this sense, the theater helped
to cultivate maturity as a set of attitudes, beliefs, and practices essential
to the process of transforming adolescent students into adult scholars,
practitioners, and bureaucrats. It played a part in a broader tale about
growing up in the early modern period. While the history of civility is
full of complexity, the civilizing role of the anatomical theater was, for a
time, suited to its didactic role. The theater served the intellectual pur-

poses of professors and the civic aspirations of the premier educational institution of the Venetian Republic. For medical students, the theater displayed the importance of their nations before a wider community. It also regulated, disciplined, and trained their physical and mental habits. By 1595, when its construction was completed, the anatomical theater reinforced the notion that anatomical inquiry was as conceptual an enterprise as it was a physical one. It helped to produce a more formal tradition of public anatomy demonstrations, encouraging auditory rather than visual apprehension of anatomical particulars. Disputation gave way to silent participation and benevolent forms of acquiescence. Aggressive, violent outbursts were restricted to the students' notes, and, perhaps in suppressed form, to their participation in private anatomies. At least, this is how the wider public viewed the situation, a perspective explored in the next chapter. In the public anatomical theater, students contemplated the natural philosophical dimensions of anatomy, listened to music, considered their own good behavior, and helped to make the enterprise and experience of anatomy a more significant part of the late sixteenth-century humanist environment of Padua.

# CHAPTER FOUR

# Medical Students and Their Corpses

I n the second half of the sixteenth century, there was a growing need for cadavers for the anatomies taking place in both public and private venues. This need not only highlighted the procedures around acquisition, but also increased the illicit traffic in corpses. It is often assumed that grave robbing served to boost the supply of cadavers, perhaps because Andreas Vesalius recommended the practice to his students. However, the legal action taken against grave robbers was severe; furthermore, there is surprisingly little evidence that it was a common feature of anatomical study in sixteenth-century Italian locales.[1] Although students in Montpellier, such as Felix Platter, described nightly excursions to cemeteries, undertaken with the help of graveyard attendants and for the purpose of finding cadavers for dissection, no Italian equivalent has surfaced.[2] Instead, authorities as well as medical students seem to have quietly followed the practice of acquiring corpses from criminal proceedings, hospitals, and other local venues.

While students were not usually out robbing graves, they sometimes acquired local rather than foreign cadavers. In doing so, they did not follow the statutes governing anatomy. Throughout the early modern period, the statutes from most Italian universities—Padua, Bologna, and Pisa, for example—provide explicit criteria for the corpses allocated to medical schools for the study of anatomy: the bodies should be low born (ignobile), and they should be unknown (ignoti), meaning nonresident or foreign.[3] In Padua, the statute was even more specific: two cadavers a year would be provided for the public anatomy, chosen from the condemned, but they were not to come from either Padua or Venice.[4] The distinction between local and foreign remained fundamental to the selection of corpses, but, faced with the need for more material to dissect, students (with the help of officials) overlooked the statutory distinction between local and foreign bodies and circumvented the regulations and oversight of clerical, municipal, and academic authorities.[5]

These deviations from the statutory and procedural norms that governed anatomical activities, and especially the problem of local corpses, augmented the suspicions around dissection. These derived from concerns about burial.[6] The Christian burial of a corpse emphasized the possibility of redemption, because the grave held the dead until the Day of Judgment. In sixteenth-century Padua, the bodies of criminals who were executed were given a Christian burial, even if their bodies were subjected to medical anatomies.[7] This represented a shift from earlier periods, when the corpse was left to rot outside the city walls as a symbol of its exclusion from the Christian community, a practice still evident in the sixteenth century in northern European locations.[8] In Italy, with the rise of confraternities in the late middle ages and a new emphasis placed on the welfare of the poor, burial rites were carefully reviewed.[9] In this context, as Katharine Park has explained, medical anatomies were troubling, both because the burial of the cadaver was delayed and because its corporeal dismemberment forced the burial ceremony to be radically altered. If this loss of honor was barely tolerated for unknown persons, who were not likely to have kin in the area and whose burial would therefore not be well attended, it was unacceptable for local persons.[10]

Because of their association with the illicit acquisition of corpses, and with disrupted burial rituals, medical students, rather than anatomists, served as magnets for the suspicions around dissection. Although scholars tend to limit the students' role in anatomical events to that of assistant or spectator, they were still fundamental to the structure and the endurance of suspicions and fears around anatomy. With a nod from the authorities, medical students were the ones responsible for the acquisition of cadavers and the burial ceremony that was supposed to conclude the annual public anatomy demonstration. The preliminary and terminal stages of an anatomical event were the most accessible to a wider public; they were the most likely place for members of the community to encounter corpses designated for medical anatomies and medical students wanting to dissect them. Community members were not invited to private anatomies, nor were they welcomed into the public anatomical theater until 1595. Thus these beginning and ending stages of an event were what reminded the broader community of students' direct association with dissected corpses.

Faced with the concerns and profound fears of the general public, on the one hand, and with the desires of the academic community, on the other, medical students engaged in a complicated series of adjustments in order to meet the often conflicting demands of these two groups. For the local populace, the medical students emphasized their commitment to the burial of dissected corpses, deflecting negative publicity around dissection and their own role in acquiring and dissecting cadavers. For the academic community, however, the acquisition of cadavers became an achievement. Medical students sought to harvest more bodies, but not (always) because they linked more corpses with the ability to learn more about anatomy, as we might think. Rather, they desired the honor that would come from their peers and professors if they found more corpses.

As this chapter will show, by participating in the study of anatomy, students learned to identify with the academic community. For medical students, this communal identity could consist of the community of foreign students that belonged to a student nation, the academic community that resided in Padua, or an imagined academic community, a republic of letters that stretched across the Continent. For example, as chapter 3 explained, the post-Vesalian history of anatomy developed a more refined public tradition of anatomical inquiry. This tradition helped medical students cultivate a contemplative response to anatomy, one that would allow them to study corpses and immerse themselves in the remains of the dead. The anatomical theater promoted this response as the *proper* response of a learned audience and of the academic community. When medical students wished to deny tradespeople entry into the theater, they argued that the presence of "gaping" craftsmen—spectators who failed to respond in the proper way to the stuff of anatomy—"diminished" the nobility of their education. By this gesture of exclusion, the students used the anatomical theater to insist on a coherent image of the academic community and their membership in that community. The anatomical theater and its public demonstrations, however, were not the only places and events for such reflection. The preparations for anatomies offered students an occasion to procure multiple cadavers, transgress statutory norms, and place themselves more securely within the academic community (in opposition to the lay community, which viewed such transgressions with horror). These activities were also part of their education in anatomy, an education that required them to look beyond their own student nation and to identify themselves more fully with the academic community in Padua, which fully supported the study of anatomy in public and private venues and which sometimes served as a metonym for the republic of letters.

## LOCAL CORPSES

While concerns about the cadaver's burial would plague any anatomical event, they appeared most urgent in situations where the corpse was thought to be of local rather than foreign origin.[11] Cases in which a local corpse was used were rarely documented, but in his chronicle of "notable things" that took place in Padua between 1562 and 1620, Niccolò Rossi included an account that renders the problem of a local corpse in highly dramatic terms:

> There was a case worthy of compassion for the cruelty of the crime, committed by the hand of Marco, the young fruit seller of 23 years, against the person of Cecelia, his wife, on 13 March [1599] while they were living in the area of Pontecorvo near S. Fregozi. Marco was young and mindful of the immodesty [*impudicitia*] of his young, beautiful wife and that many times, she may have given herself leave to be with some lover. Around three o'clock in the morning, he was persuaded by diabolical thoughts . . . called to her, and ordered that she go up to an attic which held a large quantity of fruit. Cecelia did not suspect anything of her husband, who had made her go at other times, on similar occasions. After a little time passed, Marco went away from her, and returning with a small axe, he whacked her without making any noise, and after having done it, he stripped his wife nude, killed her, and quartered her body into eight pieces.[12]

Marco was motivated to kill his wife by the belief that she had committed adultery several times. Quartering was reserved for the worst crimes, for oft-repeated ones, and for criminals that tended to occupy the lowest levels of the social hierarchy.[13] Marco had taken justice into his own hands, punishing his wife by execution and quartering. He then made a mockery of a proper burial, dispersing the fragments of her body "in the ditches of Pontecorvo and [thrown] down from the [city] walls"; her innards (*interiori*) went to the old gate and into the Brenta River.[14]

Marco's crime involved brutality, a quartered body, and a disorderly burial, all features that were routinely and explicitly associated with dissection. Marco's punishment was also deeply tied to dissection. Two days later, when the authorities elicited a confession and convicted Marco, the minister of justice executed Marco, cut off his right hand (the instrument with which the crime was committed) and draped it around his neck:

> Marco was returned to the Piazza where above an eminent gallows, he paid the penalty of his wicked misdeed before a great crowd of people for three days after the murder [*amazzamento*], and attached to the tail of a horse, his body was given to the students for an anatomy.[15]

Because Marco was a Paduan, a local citizen rather than a foreigner, his body should never have been the subject of a medical anatomy. Nonetheless, it was, and Rossi's account of Marco's case elaborates on the image of a disrupted and delayed burial in order to focus attention on the perversion of kinship relations. Marco's killing was called an *amazzamento*, from *ammazzare*, a term which means "assassination" but is often used to describe the slaughter and butchering of animals (in the slaughterhouse, or *ammazzatoio*). In the *Eleganze . . . della lingua toscana et latina* (1585), the verb *amazzare* is glossed from Orestes' murder of his mother, Clytemnestra, emphasizing the perversion of naturalized, familial relations.[16] A cognate, *mozzo-are* (to cut off), appears in *Inferno* XXVIII as a metaphor for the destruction of familial lines: Dante meets Mosca—the leader of a family feud in Florence—whose hands have been cut off and who raises his stumps, befouling his own face with blood.[17] In the same manner as these literary examples, the account of Marco's crime uses the body to represent the distortion of kinship relations. Marco's murder perverted the normal relations of husband and wife, a perversion reflected in the parodic burial of her remains. The horrific nature of the crime probably encouraged the authorities to release Marco's body to medical students.

Although the relationship between kinship, burial, and dissection has caught the attention of several scholars, discussions have been limited to the importance of statutes, executioners, city officials, and anatomists and have usually concentrated on the period before and during Vesalius's tenure.[18] Andrea Carlino, for example, has characterized the descriptions of anatomical activities, noting that the anatomist was sometimes criticized for "brutalizing, wounding, tearing to shreds, and lacerating" the corpse, thus deforming corporeal structure in moments depicted as violent and nearly sacrilegious. Dissection, he concludes, was attended by an anthropologically based concern, a superstition about handling and touching the dead, which imagined dissection as an act approaching desecration and the corpse as contaminating.[19] These notions of physical violence stem from the desecration of tombs and, more generally, from worries about burial. For the period leading up to Vesalius, Katharine Park has connected this concern to the underregulated nature of private anatomies. Private anatomies often took place in hospitals, on occasions when professors and students came across a freshly dead corpse, a body (usually of an impoverished person) that, without the protection of family to provide for a proper burial, was subjected to a postmortem dissection, even if it was the body of a local resident.[20]

These connections, and latent fears about dissection, did not disappear at the end of the sixteenth century. Rather, they migrated from anatomists to medical students. The account of Marco's crime and punishment ended not with professors conducting an anatomy, but with medical students doing so; while this was not wholly unusual, the imagery in the account permits the students' dissecting hands to echo those of the executioner as well as the murderer himself. The students, and their hands, were tainted—motivating us to investigate the nature of that stigma and the students' responses to it.

## THE ACQUISITION AND BURIAL OF THE CORPSE

The turn toward burial and religion is a recent development in the historiography on anatomy. It comes as a response to the tendency to see public anatomies as expressions of secular punishment and to frame them within historical transformations of state power (and, usually, absolutist rule). Influenced by Foucault's work on punishment, these approaches have inflated the connections between the executioner and the anatomist. As Jonathan Sawday explains, the criminal bodies used in dissection served to extend juridical power from the courts and the public squares of execution to the emerging world of science, making the anatomist complicit in the expansion of state power that characterized early modern polities.[21] To cast public anatomies solely as expressions of secular punishment, however, is to overlook the theological significance of execution and the role of public anatomies within the extended social ritual of punishment, a ritual that coupled (rather than separated) the secular and the spiritual.

As Adriano Prosperi has explained, the state had the power to execute because it controlled the life and death of the body and allocated the control of the soul—the conscience of the condemned—to clergy. Because of that allocation, "the state gained theological confirmation of its power to execute."[22] This is why, in the last days and hours of life, the condemned were assigned comforters (confortatori), who would encourage the criminal to accept the punishment of death in order to obtain divine grace in the afterlife.[23] Thus the death of the body did not necessarily imply a "second death," the death of the soul.[24] In a stunning drawing (ca. 1599) by Annibale Caracci, a comforter can be seen following a criminal up the ladder of execution; the comforter is holding a tavoletta, a small tablet with an image painted on it (figure 10). According to Samuel Edgerton's analysis, the tavoletta featured a biblical scene, meant to inspire reflection and contrition in the criminal.[25] Such props emphasize the role of redemption in this ritual, locating the individual in relation to the spiritual realm.

The scene of execution (with its punitive and redemptive associations) must also be placed within the extended social ritual around punishment, which attempted (on the bodies of the marginalized and the infamous) to provide the community with coherence and security after the traumatic experience of a crime.[26] This ritual included confession, condemnation, execution, a public anatomy, and burial. Each phase of the ritual emphasized secular and spiritual dimensions, conjoining earthly life and life ever after, condemnation and compassion, cruelty and forgiveness, sacrifice and redemption. Placed within this elaborate social ritual, a public anatomy would not be seen as sacrilegious, because questions of the soul had already been allocated to the clergy and the comfort-

FIGURE 10. Annibale Caracci, *A Hanging*, ca. 1599. Royal Library, Windsor Castle, England. Reproduced courtesy of the Queen.

ers; rather, the anatomy was both punitive and potentially redemptive. In his analysis of the case of Rodolfo di Bernabeo, convicted in Rome in 1581, Andrea Carlino underscores the attention that authorities and comforters gave to the salvation of the criminal, even as he approached the gallows. Carlino suggests that in this way, the audience could participate in the extended public ritual by accepting its punitive orientation and, at the same time, ruminating on its theological dimensions. In addition to making the criminal "pay" his debt to the community, public anatomies could inspire thoughts of sacrifice (the body) as well as redemption (the soul). This feature was exploited by artists in the iconography of anatomical illustrations, where the criminal's body was figured as a saint (or a martyr).[27]

The ritualized aspects of public executions and public anatomies help to explain why the connection between executioner and anatomist appears so fleetingly in late sixteenth-century Padua. As chapter 3 explained, Fabrici's demonstrations in the anatomical theater highlighted the civic importance of anatomy, linking the anatomist to the state by way of civic virtue rather than by executive force. He resembled a philosopher more than an executioner. This aspect reveals the contours of a new legitimacy that was created around the anatomist and his demonstration during a public anatomy, a legitimacy extending the authority of the state (via ideas about the public good) to the practices of anatomy. However, this legitimacy did not fully negate the fears surrounding dissection. Rather, those fears were transposed onto medical students, whose relationship to the state or to recognized authority was much more tenuous than that of professors. Though anatomists such as Fabrici were increasingly distanced from cadavers, medical students were increasingly connected to them, for they were seen as responsible for the acquisition of corpses and for the organization of and attendance at those burials.

Two medical students (*massarii*) were elected to oversee the preparations for and proceedings of the annual public anatomy demonstration. These students were associated with the procurement of the corpses, which they received after executions took place. They also prepared the cadavers for anatomical display during the public demonstration. Increasingly, the actual dissections for the public event were done by these students; in 1587 and again in 1595, they were renamed *anatomistae*, or anatomists, a designation that emphasizes their responsibility for preparing and dissecting the cadavers.[28] Indeed, these special students were increasingly responsible for the dissection of cadavers for the public anatomies of the late sixteenth century.[29] For example, on 28 November 1599–1600, the "anatomists [*massarios*]" were elected, individuals "who are accustomed to offer their hands as helpers in the cutting and separating of the bodies to the administrator [Fabrici] of the public anatomy."[30] Similarly, on 20 November 1600, following Fabrici's discussion of the general anatomical method and the anatomy of several animals, the "students [*massarii*]" were elected, and, "for the Excellent Anatomist [Fabrici]," they "were accustomed to offer their hands as helpers in the procuring and dissecting of cadavers."[31] These students were deeply engaged in the material practices of dissection.

In addition to handling the dead and preparing cadavers for display, these students organized and attended the burial ceremony that marked the conclusion of the anatomy demonstration and the extended social ritual of punishment. Medical students probably participated in the ceremony because a Christian burial emphasized the possibility of redemption, and because it was good publicity for the students themselves. By the mid-fifteenth century, colleges of medicine required students to pay for and attend the funeral of criminals whose corpses were used in dissection, perhaps to encourage the practice of donating bodies for dissection.[32] This requisite attendance may have prompted students to reflect on the personhood of the criminal and on the significance of burial of the dead.

The subject of death was also connected to the degree of maturity exhibited by students elsewhere within the academic community. The natural philosopher Cesare Cremonini turned to the topic of death in his lecture to university students. Echoing Aristotle's *Parts of Animals*, he claimed that students too often failed to recognize the fundamental distinction between the immutable heavens and the mutable world. They failed to learn how "Nature"—their "marvelous teacher"—"incessantly puts on and weds diverse forms" and that the "smallest of seeds" can produce the "grandest of things." Cremonini would go on to note that "in this [capacity] the oak tree is superior to gold, in this the lion is superior to the oak, and man to the lion."[33] Cremonini sought to emphasize teleological development, the idea of coming-into-being that was central to Aristotle's works. However, he compared this particular failure on the part of the students to a more common adolescent one. These students believed that they were "not born into existence from some initial time," instead thinking that they were not children and that they would not grow old; they supposed their identities to be "complete and unchanging." Cremonini ridiculed this adolescent mindset as "possible vaguely to imagine, but in no way to understand."[34] For Cremonini, maturity depended on an awareness of mortality, that is, of the nature and meaning of death. Such reasoning suggests that students were perhaps told to attend the burial of dissected corpses because it would offer them another chance to ruminate on the nature and meaning of death.

By the end of the sixteenth century, medical students seem committed to the organization of and attendance at the burial ceremonies held for dissected corpses, but that commitment sometimes served as a defensive publicity strategy. On 3 December 1597, during a public anatomy demonstration, the transalpine students emphasized the need to collect money from spectators in order to pay for the burial of the cadaver's remains. These students were following the statutes, which stipulated that the funds collected would offset the costs of the event and be given to the poor for the salvation of the dissected body.[35] By 1597, however, the Venetian Senate was subsidizing the annual demonstration. At least one transalpine student was clearly concerned about the escalation of rumors regarding dissection and about the students' lack of respect for the dead. He explained that the transalpine students wished to collect additional money because they were "deeply moved by [the] hostile rumors" that they snatched or plundered (graves) for multiple cadavers; that they profaned the dead bodies inside the anatomical theater; that, within the theater, the bodies were torn to pieces and left unburied; and that the students concealed this destruction and abomination, joining together "like dogs" to devour the corpses.[36] Here, the absence of the professor-anatomist is underscored, for these concerns are connected to medical students and to the material practices associated with anatomy. The fears around burial and about the desecration of the corpse coalesce in the figure of the medical student. The student who listed these charges claimed ignorance of such practices: "if these quarrels are not entirely empty, if they are not empty laments."[37] Responding to these charges and attempting to control the rumor mill, the transalpine students endorsed the idea of honest burial by exhorting contributions from the spectators and then attended the funeral on the 16th of December in the church of St. Sophia.[38] This public demonstration concluded with a burial; the students were motivated not only by their concern for the soul of the dead, but also by the desire to quell the hostile rumors that targeted them for their role in acquiring corpses and in dissecting them.

There is some basis for the general public's concerns, which flag private anatomies and the hazy origins of the cadavers used in them. Stories of grave robbing, although more prevalent in later historical periods, were circulating in the 1530s, especially in the context of private anatomies. In 1549, the Venetian Senate stipulated that the *podestà* of Padua should impede the taking of bodies from tombs and cemeteries for use in private anatomies; the Senate would offer 200 *lire* to anyone who turned in the delinquents.[39] In the second half of the sixteenth century, when private anatomies were more frequent, cadavers were found not (or not only) in tombs, but in local hospitals. As we saw in chapter 2, in October 1578, during their clinical rounds in the hospital of S. Francesco, medical professors decided to dissect the bodies of two women who had recently died.[40] A third woman, afraid that the same fate awaited her, interrupted the professors and their lessons. Her complaints prompted the *riformatori* to threaten the professors with the loss of their salaries and, at the beginning of the next academic year, to proclaim the statutes governing the practices of anatomy and dissection.[41] The magistrates emphasized that anatomies were to take place in January and February only, reminding everyone (lay and academic) of the appropriate time for the public anatomy demonstration. This reminder did not raise the thornier question of private anatomies, but it may be that the reminder was meant to encourage the academic community to adhere to the rules and thereby negate the need for more cadavers. Private anatomies, however, continued, even though they remained illegitimate. In the eyes of the broader community, these were the venues most likely to be used for the dissection of a local person. In the case of the above-mentioned private dissection, perhaps the *riformatori* responded so fully because one of the dead women was a Paduan.

## MITIGATING THEIR OWN FEARS: *Medical Students and Identification*

While an anatomy involving a local corpse was rarely documented, the category of "local" became highly controversial for both lay audiences (as the woman's complaint in the above hospital example suggests) and academic ones. In 1595, the year that the second permanent anatomical theater opened, the anatomy demonstration began, as usual, in early January. However, in mid-February, several transalpine students had to stand guard with rifles outside the theater. In doing so, they met with the taunts of a Milanese student: "when he began taunting them [the student guards] and our nation with crude and unworthy words, one of our boys, pointing his weapon, asks him to stop and temper his tongue from such curses."[42] The atmosphere was intensely aggressive and filled with the potential for explosive violence, but by brandishing their weapons, the transalpine students maintained a precarious state of peace. All of this excitement was related to a controversy around the identity of a cadaver that had been chosen for the anatomy demonstration.

According to the statutes, the cadavers were supposed to be low born (*ignobile*) and unknown (*ignoti*), in other words, foreign rather than local. In 1595, the first cadaver fit the bill but the second apparently did not, at least not as far as some students were concerned. The first demonstration in the new theater had taken place with great fanfare being accorded to Fabrici and the Venetian Republic. However, before the second demonstration, the transalpine students from modern-day Jülich mounted a protest because that cadaver was from their region:

> On 12 February 1595, while the student-assistants [*massarii*] were beginning the examination, a certain number of students from Jülich burst very petulantly out of the just-finished lecture of the excellent Panceroli, and, shouting repeatedly that the body they [the authorities] were releasing for the free anatomy had only just now been discovered *to be of the Germans* [italics added], [the Jülich students] try to burst into the theater with force; whom our *massarii* were resisting with inferior strength. While one of them in particular was more bold and insolent than the others, with the battle threatening even Dominus Fabricius, the whole company makes a more forceful assault so strong that their wishes are granted, with the doors nearly torn from their hinges. The student-assistants [*anatomistae*], who were working with tireless diligence to make sure the administration of the public anatomy would go forward properly and with praise, were quite distressed at this petulant and despicable arrogance.[43]

The transalpine student who wrote the account was probably not from the Jülich region, because he clearly favored the student-assistants and the use of this corpse for the public anatomy. The students from Jülich, however, realized that the cadaver was not foreign (in the sense of being unknown). They threatened Fabrici and tried to deface the new theater, nearly ripping the doors off the hinges in the same way that they expected their adversaries, the student-assistants, might rip apart and destroy the cadaver.

This episode highlights not only the physical aggression of young male students, but also the fact that, like beauty, the idea of local depended on the eye of the beholder. In this case, the Jülich students perceived the cadaver to be too familiar, that is, local rather than foreign. The violence of the Jülich students surely stemmed from their implicit recognition of the violence of dissection, for they had an insider's awareness of its procedures and its deforming effects. Moreover, their fears about the destruction of corporeal integrity were not quelled by the idea of an honest burial. Faced with this corpse, the transalpine students from Jülich were unable to accept the radically altered postdissection burial as the final stage in the extended ritual of punishment for this cadaver. While the local status of Marco, the Paduan fruit seller, was overlooked on account of his particularly hideous crime, this status was not ignored in the case of the Jülich corpse (about whose crime we know nothing). The controversy took time to resolve itself, and, in the end, the public anatomy was postponed indefinitely. The other transalpine students turned to Giulio Casseri with requests for a private lesson while the ambiguity over the precise meaning of "foreign" remained.[44] The deferral of the public event suggests that the authorities, including Fabrici, did not want the new theater to be associated with aggression or violence, or with the stigma that would come from the anatomy of a cadaver whose foreign identity was suddenly opened to interpretation. The authorities, and Fabrici, worked diligently during the subsequent decade to ensure that the theater remained "a site of tranquility."[45]

This episode is intriguing because it reveals some of the hidden divisions in the student body: a body with partitions between and within various student nations. The transalpine students experienced their studies abroad in a different manner than local students, or even those who came from elsewhere on the Italian peninsula. The transalpines were constantly reminded of their status as foreigners: local clergy voiced concerns about their Protestant leanings, while administrators and *riformatori* praised their good behavior, especially their diligence in anatomical study.[46] In this case, however, the Jülich students insisted on their difference from other transalpine students and refused to identify with their student nation, much less with the academic community. Here, identity was delineated in geographical terms, whereas earlier and elsewhere, it could be secured by religious orientation. The transalpine students were often associated with Protestant beliefs (and, perhaps because of their support for marginal teachers such as Niccolò Bucella, with Anabaptist sentiment), but this episode differentiated the students by geography (i.e., Jülich), not religion. The students from Jülich failed to identify with their nation, and the signs of failure are recorded in terms of their comportment. The Jülich students acted out; they were petulant and arrogant, shouting and forcing their way into the anatomical theater.

One way to mitigate such unrest was to make it more difficult to ask questions about the origin of the corpse, that is, its local or its foreign status. In the late sixteenth century, a new fee structure appeared for the acquisition of cadavers that allowed monetary calculations to at least partially displace concerns about the origin of any individual cadaver. In

1587, amid a dispute over whether Fabrici would offer a public anatomy, two students were chosen—rather than elected—as *massarii*. The transalpine students noted that these two were "able enough" and that their promotion was approved "on condition that they procure three bodies or cadavers, and that they ought to charge 12 shillings to anyone going to observe the anatomy."[47] The transalpines continued: "If however they should discover more than three, this would result in honor more than in their usefulness, and for these, they ought not to charge the students any more than 36 shillings."[48] In the end, "they found 4 cadavers and as payment in general took 32 shillings from each student, that is, eight for each cadaver."[49] The record does not compare this price of entry to earlier fees of three silver coins, suggesting that a new structure of pricing was emerging, one that depended on the number of cadavers and on the cost of additional burials or alms, which was the subtext for all discussions of entry fees for anatomy demonstrations.

Admitting that a surplus of cadavers would not enhance the usefulness or educational value of the demonstration, the transalpine students emphasized the connection between acquisition and honor. Equally, this remark suggests that students were beginning to perceive the public anatomy demonstration as educationally limited. The demonstration became more innovative in the early seventeenth century, when Fabrici focused on surgery. A different development can be seen in Bologna, where, according to Giovanna Ferrari, the public anatomies of the mid-seventeenth century were of little interest to both serious students and professors of medicine.[50] In Padua, public anatomies were beginning to appear useful only within severely limited parameters that were not drawn with the careful dissection of multiple cadavers in mind. Indeed, this fits well with Fabrici's habit of focusing the public anatomy demonstration on natural philosophy and on anatomy's relationship to it, rather than on the structural composition of one, two, or three corpses.

In addition, the students' comments point to the development of a new ethos in academic culture concerning the acquisition of corpses. Moving well beyond questions of educational value, this new ethic made it a mark of honor to procure multiple cadavers. The statutes allocated just two cadavers to the annual public anatomy demonstration, suggesting that sometimes the additional bodies were procured for private, not public, anatomies. From a legal standpoint, these additional corpses were illicitly acquired, but, for the academic community, the procurement of a surplus of cadavers became a sign of honor, even for student-assistants (*massarii*) as unremarkable as those described above. This was a situation that encouraged students to transgress the statutes repeatedly. It also required academic authorities to overlook those transgressions, perhaps to run interference with municipal and clerical authorities, and, when necessary, to provide protection for the students. This was a situation that produced the material circumstances for a more robust tradition of private anatomy, the subject of the next chapter.

## CONCLUSION

Everyone seems to have hesitated over the dissection of a local person. The Jülich students revolted at the prospect of attending the public anatomy of a corpse that hailed from their region. The woman in the hospital registered her complaints to the authorities over the hasty dissection of two recently deceased female patients, who were probably local. Niccolò Rossi's account of Marco's case was narrated with a rhetorically pronounced unease about the dissection of a local fruit seller, even one as monstrous as Marco. These situations raised anxieties about the academic community's adherence to the prohibition against local corpses. To counter those concerns, a tradition of public anatomy demonstrations and the construction of the public anatomical theater moved in the direction of transparency. It was there, in the open theater, that the provenance of the corpse was meant to be publicly revealed. In 1605, there is an admittedly strange case in which a local man (*civis patavinus*) came to Fabrici's public demonstration inside the theater, because he was asked to do so by the widow of a man whose body was one of the cadavers.[51] There is nothing in the record to suggest that the dead man was a Paduan. Rather, the case implies that the provenance and the details of this corpse were widely known—transparent to the widow as well as her friends.

The same transparency is not evident in the private anatomies of this period. These remained a source of concern, because they used corpses that were illicitly acquired and probably were often local rather than foreign. While university statutes made the acquisition of local cadavers illegal, this stipulation was sometimes overlooked in practice. In the very few documented cases that involved the dissection of local corpses, their depictions of medical students are accompanied by a range of aberrant expressions and latent fears about dissection. Burial remained the central motif in these expressions, but it is worth noting that the dissection of a local person made it more difficult to pull off an honorable burial—because it would be attended by family members who would feel dishonored by the delays occasioned by the dissection and by the way it altered the corpse—and made it more likely that attendees would voice concerns like those of the old woman in the hospital and the Jülich students.

Medical students played many roles in the history of anatomy in the post-Vesalian era. While the next chapter will discuss their experiences and the educational merits of private anatomies, this chapter has explored the inconsistencies in their approaches to dead bodies. Even though medical students were staging their own commitment to the burial ceremonies of corpses dissected in the public anatomical theater, they also devoted their energies to the acquisition of multiple cadavers for private anatomies. Rumors circulated, emphasizing the violence of dissection and the disruption of burial rites and graves, and accusing these students of devouring cadavers, leaving the remains unburied, or, worse, trying to hide their "abominations" (just as Marco did). Medical students—by transgressing the statutes and, thus, the tacit assumptions upon which public trust was granted for medical dissections—also participated in an academic culture that would award them honor if they procured a surplus of dead bodies. Moreover, this ethos appealed to the transalpine students, who had a reputation for and a documented history of adhering to rules. Although these students were already scrutinized for their Protestant leanings, even they could be persuaded to choose honor and fame above lawfulness when it came to the study of anatomy and their medical education.

If crime is an event in which a culture fails on its own terms, "a moment when microsystems challenge macrosystems of power and values," as Edward Muir and Guido Ruggiero suggest, then the crimes discussed in this chapter are particularly telling.[52] Historians look to crime, Muir and Ruggiero continue, in many ways: "breaking a law sets apart certain actions as a special kind of event, which can disrupt some social solidarities and create others; crimes exercise power in ways that reveal social hierarchies and fields of disempowerment; they communicate values, especially contested ones; and they generate symbols for social relationships."[53] Marco's crime was to murder his wife, that is, to take justice into his own hands and execute her for the crime he thought she had committed. It highlights the gendered hierarchy of marriage and an entire field of disempowerment. The illicit acquisition of corpses (especially local ones) also constituted a crime, but this form of crime was not set off as a special kind of event. Rather, it became part of a medical student's education. It created a "social solidarity" among students (and faculty—the anatomists offering private anatomies did not reject corpses); and it communicated values about the importance of anatomical study for physicians and about the untouchable position from which noble, educated men viewed the rest of society. For some, the stigma associated with these crimes was transformed into a mark of honor, thereby laying the foundation for a stronger, more vibrant tradition of private anatomies in the post-Vesalian era.

# Private Anatomies and the Delights of Technical Expertise

In the winter of 1598, Padua was full of anatomical activity and the hint of a generational rivalry between the eminent and aging anatomist Girolamo Fabrici and his former student, the experienced surgeon and anatomist Giulio Casseri.[1] Two transalpine students, conciliator Ioanne Svenzelio and proconcilator Abraham Haunoltho, recounted the proceedings, first describing Fabrici's public anatomy demonstration:

> At the beginning of February, he [Fabrici] himself speaks about the universal aspects [*universali*] . . . not about the particular and the more exact parts [*particulari et exactiori*], which he has drawn out in this year all the way up to the 12th of March concerning the organs of apprehension, namely the organs of movement [*progressionis*] and respiration.[2]

Here is further confirmation of my argument that, in the hands of Fabrici, the public anatomy demonstration became a vehicle for the philosophical presentation of anatomy, one that took up the universal aspects of anatomy—the Galenic category of usefulness—but often ignored any immediate connection to the details that may have been visible in the dissected corpse, lying prostrate on the table in the center of the anatomical theater.

Although medical students had come to expect this from Fabrici (and from the public demonstration), they were especially quiet in accepting it in 1598, because they encountered more exact treatments of anatomy in other venues. Not only did Fabrici offer these details in a lesson that extended beyond the public anatomy demonstration, but Casseri also concentrated his attention on the particular structures and procedures of anatomical inquiry. Indeed, students especially eager to learn about the details of anatomy from corpses and by way of the techniques of dissection and vivisection rushed ahead to Casseri's private anatomy, which left no "obscurities" and met with "great applause": it covered anatomical topics in a monkey, the vivisection of a dog, and the dissection of nine cadavers.[3] If private anatomies were usually improvised (when cadavers became available)—done on the fly, perhaps hastily or haphazardly, and often in secrecy—Casseri's demonstration reveals the strength and vibrancy of a tradition in the making. In 1598, more resources went to the private anatomy demonstration than to the public one. Not only did Casseri dissect and discuss the anatomy of nine cadavers (a number never before reported for public or private anatomies), but his demonstration lasted five weeks, the amount of time usually allotted to public, not private, anatomies. Moreover, five weeks would have been a sufficient amount of time for authorities to shut down the event if it was seen as illicit and controversial. That the officials did not do so raises the possibility that private anatomies, under certain circumstances, were tolerated and on their way to becoming legitimate venues of study.

As the students' account reveals, the terms of anatomical inquiry were being reoriented around the different goals and contents of public and private anatomies. At the heart of this reorientation was the uncertain supply of corpses, partially addressed through a new ethos around acquisition that permeated academic culture and, especially, its roving bands of students (see chapter 4). Seeking honor and appreciation, medical students undertook a search for numerous cadavers, instead of just the two bodies the statutes required them to supply for the annual public demonstration. While corpses were always difficult to find, the beginning of the seventeenth century is notable for the bounty it produced. This chapter investigates the use of those cadavers in private anatomies by tracing the growing opposition between the public and private traditions of anatomical inquiry as well as a notable interest in surgery. Not only was Casseri's private anatomy devoted to exactitude and the uncovering of particular structures, but it also emphasized technical skill—the skill to vivisect a dog and to dissect nine corpses so that, in the dog and in the corpses, motions could be detected and the specific bodily structures could be seen.[4] Whether the cadavers were dissected consecutively by Casseri or simultaneously by Casseri and his students, the presence of nine corpses suggests that manual techniques were central to the lesson, not only to what students learned, but also to how they learned: by observation, by imitation, and by trial and error. Unlike the anatomical theater, the spaces where private anatomies were held were compact and capable of supporting a technical focus. This reduced space made it possible for students to learn how to flay and dissect corpses and vivisect animals, in part because it brought students, professors, corpses, and animals closer together. The smaller spatial dimensions should not be overlooked in explanations of how and why private venues of anatomy were consistently depicted in the late sixteenth and early seventeenth centuries as spaces for observing, that is, as spaces where students *could* see.

Of course, additional explanations are needed to account for the interest that these private anatomies commanded and to understand their relationship to public anatomies. As an anatomist and a teacher, Casseri aimed to have the students learn to manipulate specimens, inspect their parts, and contemplate their uses. This threefold nature of anatomical learning was derived from Galen, for whom anatomical knowledge came from descriptions of anatomical structure (*historia*), from an understanding of particular functions in relation to structure (*actio*), and from a more contemplative assessment of the usefulness of the parts or the process, which expressed a universal principle (*utilita*). Although these were the ideal features of Galenic anatomical pedagogy, Casseri emphasized the need to explore structures and functions by "manipulating" and "inspecting" the parts of the body before "contemplating" their uses. The first two kinds of engagement focused attention on the object (either corpse or animal) and on the skillful dissection performed by the anatomist, an attention that was powerful enough to transform the third kind of engagement, contemplation, into an activity that depended more fully on what was separated by touch and distinguished by sight. Casseri's emphases bore the traces of a tradition of anatomical inquiry that took place in private, not public, venues.

These emphases and features of Casseri's program, however, are fairly abstract. They tell us what might have happened in private anatomies, but they indicate little about what brought students there in the first place. Most medical students, after all, were not anatomists-in-training. For decades, medical students had articulated a deep dissatisfaction with Fabrici's teaching, which was often narrowly focused or confusingly delivered. This surely pushed them toward alternative lessons, such as private anatomies, and alternative teachers, such as Casseri. Nevertheless, the robust nature of the private anatomies at the end of the century also depended on changes in anatomical instruction that were specific to the last decade of the sixteenth century.

Medical students sought out and attended private anatomies because these private events offered exclusive experiences that covered general topics in anatomy and surgery. In 1595, when the public anatomy demonstration took place in the permanent anatomical theater, some students were troubled by the public status of the event, which no longer was limited to the broader academic community, but instead was offered to both academic and nonacademic audiences. The anatomical theater mixed social orders, and thus sent some students running to the private venues of anatomical study, where their privileges were secure and their status could be reconfirmed. There, they encountered general instruction on anatomy and surgery, a function served many decades earlier by the public anatomy demonstration. Moreover, while both public and private anatomy demonstrations incorporated instruction on surgery, the surgical lessons in public tended to cover a few surgical operations and offer a display of instruments, without incorporating any anatomical study (beyond what was necessary to understand the discussions of surgery). In the private anatomies of Casseri, however, instruments were used to model the motion and function of muscles, that is, to promote an understanding of morphology and physiology.[5]

Despite recent interest in the role of craft and in technical traditions of knowledge during the Scientific Revolution, in histories of anatomy the significance of surgery has been limited to the background experience of early anatomists, particularly Berengario da Carpi, and to the role of the surgeon in the public anatomy demonstration. As we saw in chapter 1, the public demonstration was organized around a *lector*, an *ostensor*, and an *incisor*, and it was the last role, the dissector, that was usually played by a surgeon. However, this structure was only maintained in public demonstrations (and even there, as the foregoing chapters have shown, it was not always followed). In private anatomies during the late sixteenth century, the anatomists dissected, displayed their technical skill, and promoted the joint study of anatomy and surgery. According to their students, these feats were done with enthusiasm, carefulness, and delicate hands, and the results were exquisite.

By treating aspects of surgery and focusing attention on the objects of study, private anatomies developed into a tradition of anatomical inquiry that gained luster and fame as "the single most significant center for anatomical research and teaching" by the end of the seventeenth century.[6] This constellation of features and practices has gone undetected by scholars who tend to view the work of Andreas Vesalius as prologue to the work of William Harvey. Rather than connecting Harvey to Vesalius (or to Fabrici), Harvey should be seen in light of the developing tradition of private anatomies, which directed attention to the technical expertise of academic anatomists. This technical focus reveals a crucial link between this era and the Enlightenment that succeeded it, for the ways in which students and anatomists behaved and operated in private anatomies provides a prehistory to the forms of attention that dominated Enlightenment inquiries into nature. In the Enlightenment, the fixed attention of naturalists served to elevate the mundane into an aesthetically pleasing object *and* one appropriate for sustained scrutiny and study.[7] The private anatomies of the early seventeenth century featured decaying corpses, transformed into objects of study and scrutiny; even if only occasionally beautiful, the dissections by expert anatomists nevertheless were noted for their carefulness and their clarity.[8] These private anatomies introduced a new visual focus on anatomical parts and instruments of intervention; these were combined in a pedagogy that sought proximity between teacher, student, corpse, and animal and fostered a deeper appreciation of technical expertise within the academy.

## THE VIRTUES OF PRIVACY

Private anatomies contrasted with the open public anatomies in the anatomical theater in several ways, but students understood the "private" nature of private anatomies in terms of exclusivity. In 1595, when the second anatomical theater opened, Fabrici had written a letter to the Ve-netian Senate requesting that in subsequent years, this theater remain "open" and entrance to it be free of charge. Several medical students were angered by the large crowds at the public demonstration—and especially by the presence of craftsmen and tradesmen (*meccanici*)—that prevented students from savoring the "sweetest fruit of anatomy."[9] On the pages of the transalpine nation's official records, their students' anger culminated in a rhetorically rich, violent, and futile outburst against the openness of the theater and the inclusivity of the public demonstration:

> I counsel that there should be no credence given to the opinion of those who try to introduce anatomy for free in this academic institution, a sort of disease of the humanities, and the most open window of sedition and murder, contrary to our ancient traditions that have been preserved all the way to the present age. Indeed, what would agitate more vehemently or excite to arms more quickly those noble and learned members of the most famous school, those enjoying the extraordinary and unassailable privi-leges, than to witness with a hostile eye or to allow a crowd of the worst sort of gaping craftsmen and vulgar men to occupy the lower benches—nay! the whole theater—to trample, diminish and impede that sweetest fruit of anatomy?[10]

The students made specific mention of the ancient traditions of learn-ing because the study of anatomy had been imbued with broader cul-tural significance. The permanent anatomical theater was a catalyst for this remarkable investiture. In a moment of outrage, these students in-sisted on the humanist foundation of their education in anatomy and on an image of the academy that was exclusive: an institution, a pedigree, and a set of habits reserved for the noble orders of society. The presence of craftsmen in the anatomical theater had struck a nerve. Future anato-mies were uncertain—would bodies be made available? would space be found to demonstrate? would Fabrici uphold his responsibilities as a teacher?—but the students' remarks extend that uncertainty to human-ist education, which, at the end of the sixteenth century, looked to them as if it was in a state of decline.

Students experienced what they considered the contaminating pres-ence of the craftsmen in the theater. They longed for an exclusive space, one that denied entry to anyone not affiliated with the university, and the terms of their desire point to what they found in private anatomies. In

the public demonstration, craftsmen "gaped" at the anatomy and ob-
structed the views of the students, implying that in private, without such
obstructions, students were free to look (or perhaps even gaze) at the anatomy that was underway. Concluding their protest, the students recon-
nected themselves to the scene of the public demonstration:

> Let it remain hated, if we are wise, let this incitement to idleness remain
> hated, since it would be better in the present circumstances to use even a
> slippery knife, however feeble, than it would be to enjoy fully an uncertain
> hope in future circumstances [for our education]."[11]

The students had been told to settle down. They had been "incited to idleness" and thus could only defend their honor and their education with
words, not weapons. They had to rely on violent words rather than deeds,
though they wished to take up weapons, even feeble ones. The metaphor
of "slippery handle (ansa lubrica)" (a notable hyperbaton in Latin) emphasizes the students' connection to the scene. It was a public anatomy demonstration, but they felt an affinity with the most active students at the
event, the student-assistants (massarii), who were again being called anatomistae, or anatomists.[12] It was these students who took up knives, even
slippery ones; their technical skill was meant to be seen as virtuous and,
in the rhetorical construction of the passage, their personae as violent and
virile. This was the basis for the metaphor and, however obscure, a clue to
this demonstration. The transalpine students longed for the exclusivity of
private venues, for unobstructed views of anatomy, and for the chance to
gain (not only to witness) technical skill for themselves.

Students considered the practice of dissection to be a worthwhile
technical activity, but they refused to see it as an activity that connected
them to craftsmen. Instead, within learned medical traditions, technical
skills and technical expertise were cultivated as extensions of learned
surgery and as distinct from the abilities of craftsmen, including barber-
surgeons (barbitonsores), empirics, and butchers. Professors and students
strove to exclude residual elements of craft, such as the craftsmen who
had come to the anatomical theater, and to give technical expertise an
exclusive character. Students characterized dissection as a privilege, one
not given to every student. Technical skill in students became a sign
of accomplishment and educational progress. A letter from June 1597
praised the student Johannes Richter for pursuing practical medicine,
debating the arts of medicine, and "devoting himself to the laborious

work for the administrations of anatomy and surgery."[13] The students' fascination with the manual activities, or "laborious work," associated with anatomy, dissection, and surgery could produce surprisingly fervent rhetoric. The author of this letter, Richter's admirer, elaborated on the topos of devotion by noting that through Richter's diligent study and labors, "he acquired fame for his talent and learning, which caused the eyes of all to turn on him" and transformed him into "an ornament of the entire transalpine nation."[14] Similarly, in a letter from January 1598, Jacobus-Israel Brunner is praised for being versed in Hippocratic learning and for his skill with various surgical operations and anatomical exercises.[15] Brunner went on to become procurator in 1598.[16] These letters attribute value to surgical knowledge and to the technical skill associated with surgical operations and the labors of anatomy. This technical expertise, displayed more fully by professors such as Casseri, could be assimilated into a university education because it was sharply differentiated from the skills of craftsmen and was assigned importance within the fields of anatomy and surgery. In this, both public and private anatomies played a role.

## FABRICI AND LEARNED SURGERY

While anatomy demonstrations, especially private ones, had always treated aspects of surgery, it was not until the late sixteenth century that surgery became both an object and a tool for research. Public anatomy demonstrations encouraged a stronger connection between anatomy, surgery, and medicine and between Padua and Venice. According to the medical curriculum at Padua, surgery demonstrations had been on the books for a long time. The reforms of the 1580s discussed public anatomy demonstrations, surgery demonstrations, and private anatomies, but throughout the 1580s and 1590s, Paolo Galeotto and Giulio Casseri provided instruction in surgery during their private lessons and anatomical exercises. While Fabrici was responsible for demonstrating anatomy and surgery, the records of the transalpine students indicate that he only began to pursue the latter with any frequency (and consistency) when, in the late 1590s, he took up surgery as a research domain.

Inside the anatomical theater, Fabrici treated aspects of surgery and, in this public venue, connected himself fully to textual traditions: to the humanist reappraisals of surgery from the classical, Byzantine,

and medieval periods; and to the contemporary culture of print on surgery in nearby Venice. In 1596, without a satisfactory cadaver for the public demonstration, Fabrici lectured on two Galenic works of surgery—*De
Tumoribus praeternaturalibus* and *De Articulationibus.*[17] In that same year,
he wrote a brief letter of praise for the second edition of Giovanni Andrea
della Croce's (1515?–1575) *Chirurgiae universalis opus absolutum* (1596),
published in Venice. Fabrici's letter called attention to the illustrated collection, or catalog, of surgical instruments contained in the book.[18] In
November 1599–1600, he himself displayed a collection of instruments
inside the anatomical theater:

> At the end of November, I, with the procurators of our nation, D. Raab
> and D. Dietero, approached the Excellent Aquapendente, about what we
> should expect for this year about a possible anatomy demonstration; also,
> as a favor to the transalpine students, so many of whom had gathered to-
> gether, we asked that he choose to present in their midst administrations
> of anatomies in a timely fashion and to offer something extraordinary and
> worth seeing; what he promised to do was the showing of the lower ven-
> ter, as well as surgical operations, and that they would surely see the most
> complete exhibition of surgical instruments.[19]

Fabrici was praised for the complete exhibition of instruments, though
the student's account suggests that the spectacular aspects of the surgery demonstration culminated in the display of instruments. In this
respect, Fabrici's demonstrations differed radically from Casseri's use of
instruments. Casseri used instruments to model motion and thus investigate the function of sets of anatomical parts and the processes of motion and speech.

Even when Fabrici covered surgical operations in his demonstrations,
he kept anatomy and surgery separate.[20] On 22 January 1600–1601, after
the assistants, or "anatomists," procured the body and had begun the
vivisection of animals and the dissection of the corpse, Fabrici again
turned to surgery: "resuming at least some of his vigor, the most skilled
man hastened to the demonstration of surgical operations; he exhibited
with such faith and skill."[21] Even though his health was failing, Fabrici
provided several notably accomplished demonstrations, but the description emphasizes the student-assistants as the real anatomists, procuring
bodies and dissecting them. Similarly, in October 1602, "on behalf of our
Nation, I approached the Excellent Aquapendente, I had come to learn of

what hope there would be of an anatomical study . . . he promised above all a most exact anatomy, as soon as his former state of health would be restored."[22] By March, Fabrici had recovered:

> In the theater he began well, no one pressed him, and no one pushed against him, but was lead by the love of his studies, which had blazed from within him this year. He longed indeed to cover all the demonstrations of surgical operations, which he presented as being more profitable in this long anatomy, but he was dissuaded by the noise of the students, who demanded the anatomy, which . . . he provided.[23]

Fabrici wished to present surgical operations (all of them) and, unlike his earlier demonstrations, he listened to the interruptions and demands of students. Rather than joining anatomy with surgery, using one to examine the other, he turned back to questions of anatomy.

Fabrici's interest in surgery can be detected in the books of his personal library.[24] This interest was also widely recognized. Late in 1601, the rectors wrote to the *riformatori*, pressuring them to act on a request for private anatomies. The letter noted that "on all the days, that is on the ordinary ones and extraordinary ones, assiduously until Easter, not only does he [Fabrici] make the anatomy, but he demonstrates all the surgery operations, showing how one can treat wounds . . . comparing ancient with modern ways, and really brought together all these things, like a good surgeon, with great effect and profit to the students, and universal satisfaction . . . which had not been seen for many years or maybe ever in this school."[25] The letter highlights the continued significance of the anatomical theater as a demonstration space. Unlike Bologna's anatomical theater, Padua's theater was not made obsolete as an educational venue.[26] Rather, it brought attention to the topic of surgery and, in the absence of a sufficient number of cadavers, it would have been the only working venue. Moreover, in the early sixteenth century, new editions of classical, Byzantine, and medieval surgery texts began to appear, generating a humanist reappraisal of the field. Fabrici's decision to cover aspects of surgery inside the theater—and to flag this reappraisal—invites us to reconsider the relationship between the university and its nearby urban environs.

An eminent professor's interest in surgery may seem as curious as that of his students if we imagine that physicians saw themselves as superior to surgeons and disconnected from surgical practice. In urban settings such as Venice, however, that hierarchy was less pronounced.[27] Scholars have convincingly shown that the *medico chirurgo*, or "graduate surgeon," was closely associated with the Renaissance physician, with colleges of medicine and surgery, with apprenticeships and licensing examinations, and with patient care.[28] In sixteenth-century Venice, physicians and surgeons were publishing widely in Latin and in the vernacular; they practiced together, sharing patients as well as thoughts on diagnoses and treatments. These men shared networks of associations that ran throughout the Venetian Republic, serving on military expeditions and public health boards; their interests and their commitments were part of the medical profession as it took shape in the late sixteenth century. Fabrici registered the significance of this tradition, engaging it directly with his own publication in 1604 as well as with his teaching.

Historically, the university prepared a medical practitioner using material drawn from medicine, or *physic*, as well as surgery. At Italian universities, including Padua, students could earn a doctorate in surgery; though not many did, most medical students had at least some experience with surgery.[29] In contrast, in the north, degrees in surgery were not possible, so the surgeon was consistently aligned with lower trades and with barbers. In Italy, surgery was not limited to manual skill; it also included knowledge of the theoretical principles underlying medicine, the practices of textual commentary, and specific operations—all features present in Fabrici's publications and much of his teaching on surgery.[30] These aspects would allow medical students to follow the professional trajectory that, in the future, would place them alongside experienced physicians and surgeons.

Students also cultivated their knowledge of surgery. Not only did they prepare and dissect specimens for Fabrici's public anatomies and watch appreciatively as Casseri dissected expertly in private anatomies, but they also established stronger ties with local practitioners. By the end of the sixteenth century, medical students in Padua were accustomed to interacting with learned practitioners in Padua and, especially, in Venice. On 9 January 1597–1598, the transalpine students noted the death of Doctor Tiberius Phialetus, the surgeon and anatomist from Bologna, who died of paralysis.[31] On 21 April 1599, these students gave two gifts from their treasury to acknowledge the death of the palsied surgeon, Johann Bleuvfuss Hasso.[32] In late November 1606–1607, the transalpine students were appreciative of Adriano Spigelio (1578–1625), a *medicus ordinarius* who was connected with the teaching of medicine for their nation and was the author of books on herbals and on anatomy. Spigelio donated a copy of his book on herbals to the transalpine student library, a gift that recognized what appears to be a highly compatible and mutually rewarding relationship between the professor-practitioner and this group of especially eager students.[33] In this same year, amid the transalpines' concern about the relations between their members, local apothecaries, and tonsors (*barbitonsores*, or barbers), they wrote of the death of one of their own, Francis, a surgeon and *medicus ordinarius*.[34] These students' interest in surgery continued, for in 1610 the transalpine library acquired a copy of the vernacular edition of della Croce's *Universal Surgery*, one of the texts that had fascinated Fabrici at the beginning of the decade.[35] In the face of book exchanges, the transfer of knowledge, and new topics treated inside the anatomical theater, the boundaries between the world of study at the university and the professional world in Venice appear remarkably porous, suggesting that the academic traditions of medicine were responsive to the shifting demands being made on learned practitioners.

## TECHNICAL EXPERTISE

Private anatomies provided some of the basis for that subsequent collaboration between physicians and surgeons. In contrast to public anatomies, which developed the humanist study of surgery, private anatomies focused attention on technical skill. In 1595, for example, Casseri gave a private demonstration and dissection that the students found extraordinary. After Casseri discussed the topic of generation (as Fabrici might well have done), "two cadavers were provided for him, a male and female, on which he administered a perfect and complete anatomy, but also he demonstrated earnestly and very carefully surgical operations" and received "the highest praise."[36] Although this private anatomical exercise began with a discussion, or lecture, just as a public one might have, it featured the activities of dissection and surgical operations, elements that clearly distinguished it from public anatomies. Here, the dissections were called "complete" because they covered the whole body; and the surgical operations were carried out "very carefully." The phrase is *studio accuratissime* (from *adcurare*), an activity done with studied care (or done carefully), and it joins carefulness with exactness (applied to objects rather than to individuals, for whom the term "diligence" is used). While private anatomies continued to provide a comprehensive treatment of anatomy (the whole, and the parts in relation to the whole), they also encouraged an anatomical pedagogy that focused attention on the manual techniques of dissection, vivisection, and surgical operations.

By the early seventeenth century, medical students eagerly pursued lessons on surgery because they wanted to learn how to perform operations. This desire magnified the importance of technical skill. Private anatomies, as the place for that instruction, began to be seen as important educational and research venues and, on that basis, were cultivated as a distinct anatomical tradition. At least as early as 1601, letters passed between the *riformatori* and the rectors of the university that spoke of the need for private anatomies: "seeing that anatomy is more useful and necessary to medicine and because so many have need to see one and to see it again assiduously [*et rivederla assiduamente*] and not only the particular students of the transalpine nations but many other students."[37] The letter dated 18 July 1601 noted that Fabrici's lessons were carried out "with every bit of diligence, exactness, and learning [*diligenza, esquisitezza, et dottrina*]." It then requested "humbly" that the Republic concede that during the vacation period—which, according to the statutes, was the time allotted for public anatomy—"the students be able to have a private anatomy" in order "to try out that which they have learned in public at the lessons of Aquapendente, and in a brief time to learn of this art [*in più breve tempo imparar de quest'arte*] that is so necessary to Medicine."[38] This letter attests to the need for private anatomies, and it also highlights the technical training that the private anatomies sought to encourage. In private, students learned the "art" of dissection.

Learning this art required corpses, and both the rectors and the *riformatori* recognized that need and worried over it. In another letter from 1601, one of the rectors explained that "to make these operations, they [the students, and perhaps the instructors] need many corpses . . . and because some few students are being guided by a certain Doctor Piacentino [Casseri], who does not have a seat at the Studio, they bring much disturbance and obstructions, they steal bodies, as if instigated [by this need]." "Believe me," he wrote, "it would be advantageous to offer at the end of the public Anatomy the ability to make private ones [*particolari*] with permission and stewardship over them from the Rectors." The *rifor-*

*matori* denied the request on account of disturbances, and they reminded the rectors of the criminal repercussions for transgressing the decrees of the magistrate (presumably the decrees requiring bodies to be foreign, and perhaps those against disturbing graves).[39] The rector's letter construes the relationship between public and private anatomies as an opposition, emphasizing the private venue as a space for students to practice technical arts and for the anatomist to display his expertise.

The institution's ambivalence over private anatomies continued. On 12 February 1604, when Fabrici was severely ill and had taken to his bed, the administration gave Casseri responsibility for the annual demonstration: "it was adopted cheerfully," but rather than offer a public demonstration, Casseri "exhibited to us a perfect and comprehensive [*perfectam ac sufficientem*] anatomy in his house."[40] He may have decided to conduct the anatomy privately, out of respect for his aging colleague. In the following year (1605), even though Casseri had been granted most of the responsibility for the annual public anatomy demonstration, it was decided that "only for Casseri, [temporary] permission would be given for private anatomies."[41] In 1613, in addition to ongoing arguments around a special figure called an anatomical counselor,[42] there was an attempt to grant private anatomies legal standing, that is, to set a precedent that reflected their growing significance and their separation from the public anatomical tradition.[43] Medical students and administrators began to consider "whether a license for having and promoting private anatomies seemed able to be obtained from the most Serene Venice without legal permission."[44]

### PRIVATE ANATOMIES

Originating in his own surgical practice at the hospital of S. Francesco and cultivated while he was a student of Fabrici, Casseri's commitment to technical skill and to providing technical instruction was developed in the private anatomies of the late sixteenth and early seventeenth century. Technical skill is an attribute reflected in his author portrait (figure 11), which depicts Casseri in the act of completing a dissection of a hand. The image draws upon an iconographical tradition of representing the anatomist at his work, but, unlike Vesalius's portrait (figure 12)—which emphasizes the physicality of dissection alongside his own and his cadaver's muscular shoulder and forearm, for example—that of Casseri

FIGURE 11. Giulio Casseri, *De vocis auditusque organis historia anatomica* (Ferrara: Victorius Baldinus, 1601), author portrait. Reproduced courtesy of the Boston Medical Library in the Francis A. Countway Library of Medicine, Harvard University.

FIGURE 12. Andreas Vesalius, *De humani corporis fabrica* (Basle: Oporinus, 1543), author portrait. Digital image, M0008926. Reproduced courtesy of the Wellcome Library for the History of Medicine, London.

emphasizes the hand in the absence of well-defined muscles. Casseri's portrait, as Elisabetta Cunsolo has explained, dramatically renders the delicacy of the operation and the hand, a significance further elaborated on in the playful inscription featured beneath the portrait: *Rimatur manus apta manus: mens erue mentem*, "the skillful hand lays bare the hands;

O mind, lay bare the mind."[45] Based on this presentation, Casseri defined himself as an anatomist by virtue of his technical expertise, claiming a knowledge that derived from his skilled hands.

In Casseri's texts and in records of his instruction, the practice of and procedures surrounding vivisection became important indicators of manual expertise. In 1613–1614, Casseri conducted a private anatomy that concluded with the vivisection of a dog and a display of the recurrent laryngeal nerves (once severed, the dog ceases to bark): "At which time after he had demonstrated the recurrent nerves in a dissection of a living dog, turning to the spectators he thanked them, because they had hitherto deemed it worthy to listen to him teaching with such great interest."[46] Basically, he turned and took a bow. The vivisection of a dog had become a convention in anatomy demonstrations—an especially dramatic one that called attention to space and to the distance between the teacher, the audience, and the dog. Vesalius vivisected a dog in Bologna in 1540, only to have the Italian students mangle its heart when they came up to the table to see it. Colombo also vivisected dogs, in order to expose the workings of the laryngeal nerves, and his descriptions emphasize the great crowds present and how their presence made it difficult to see the motions of the animal that his careful vivisection had exposed to view. Casseri's demonstration concluded with and culminated in vivisection. The climax, while full of showmanship, enabled a more personal transaction to occur between him and his audience. After Casseri thanked them for their support and appreciation (thus acknowledging the pleasure they took in the event), he asked, "as if as a favor," that the spectators "attend the remains [*exuvias*, the spoils] of the dissected body at the appointed shrine in the burial ground of the Servati."[47]

Casseri echoed Colombo's chapter on vivisection and made vivisection a central feature of his own methodology. At the beginning of his publication on speech, Casseri articulated a composite methodology for the study of anatomy. The first part was called an "applied and operative" *historia*: "it unfolds the body's composition [*fabricam*] with great accuracy and produces so detailed an acquaintance with even the smallest parts that from this skillful dissection, the parts, whole and unhurt, can one by one be separated off even on a living body."[48] The anatomist's attention to structural detail was nothing new, but Casseri highlighted the technical skill necessary to "unfold" the body and leave the separated parts "whole and unhurt . . . even on a living body." While that skill was routinely displayed in a dissected body, it was tested and proven on a living one. Vivisection was a procedure for the expert.

Casseri's methodology developed in several ways. He was a former student of Fabrici's; he assisted in public demonstrations; and he worked as a surgeon at the hospital of S. Francesco. With his academic training and his practical experience, Casseri promoted a methodology that retained elements from Fabrici's instruction and research: the category of usefulness and the contemplation of more general principles. However, these elements were also shaped by Casseri's practical training. In the dedication to Rainutio Farnesio, the Duke of Parma, at the beginning of Casseri's work on the senses, Casseri stated that he would "inspect the secrets and mysteries of nature [arcanaque mysteria introspexisset ]"; in a letter to the reader, he echoed this idea, stating that he would "inspect the divine construction of the human body and contemplate it [quaedivi- nam humani corporis fabricam inspicit, et contemplatur]."[49] While the sentiments had become a standard part of the anatomical tradition, contemplation, in Casseri's private anatomies, depended to a greater extent on an inspection of the body.

Casseri, even more than Fabrici, sought to transform contemplation into a visual experience. He wished to emphasize not only the ability to hold an object in the mind's eye, but also the ability to sustain the scrutiny of an object held before one's eyes. He explained that his knowledge and his ability to write a treatise on the senses depended both on long experience and on the things he would observe by contemplation (con- templatione observassem), suggesting that the word "contemplation" identified a posture of sustained viewing.[50] Fabrici had reserved this term for the final stages of his project: one contemplated the axioms of anatomy, that is, its universal character. With Casseri, however, contemplation was a form of engagement with visible phenomena, a posture of sustained viewing, or what we might identify as staring (at an object rather than off in space).

Following the patterns in classical works on philosophy and anatomy, Casseri proposed to study the human body by combining technical skill (peritia) and contemplation, that is, "by joining expert examination to diligent contemplation."[51] Casseri recommended a change in what was then the current pedagogy: unlike other anatomists who proceed from the parts to the whole (beginning with a description of those parts in the phase called *historia*), "I proceed by the opposite and, as it were, free method, from the whole to the parts."[52] This was a clear contrast to Fabrici's work. Fabrici used the image of a *theatrum* to emphasize the Whole Animal, which was then broken down according to the processes of the organic soul (rather than according to specific anatomical parts, as was the case with Casseri's illustrations and methodology). This new direction meant that the focus, as Casseri emphasized, was on the parts. In one of the tables in his book, the central image of the neck and head of a male cadaver represents the whole around which the detached parts appear (figure 13).[53] Casseri imagined these anatomical tables simply as an initiation. To engage more fully in anatomical study, he recommended that a student remove "hand-drawn tables"; once they were no longer available, the student would begin to observe and contemplate structure.[54] Here, it is worth emphasizing the point that Casseri was still thinking about a posture of sustained viewing, but he wished to substitute the object that was being viewed—going from picture to corpse. The process required a student to focus on the dissected parts that lay before him, distinguishing them visually according to what he remembered from the image.

As did many of the anatomists who preceded him, Casseri followed Galen, noting that the first part of his own method was the true anatomical method, the one Galen treated in the first book and the first passage of *De usu partium*. Casseri also added a second feature to his method, a theoretical one: "on the basis of inherent constitutions, consequences, and accidents, it explores the uses of the parts and examines their points of usefulness by means of mental acuity alone, actual dissection being bypassed . . . which Galen also mentioned." Such comments, while still well within the Galenic fold of anatomical inquiry, help to place Casseri in the context of post-Vesalian anatomy in Padua. Casseri studied and worked under Fabrici, who, in his publications and in the classroom, explored the relationship between anatomy and natural philosophy. From him, Casseri learned about usefulness, the most theoretical category in Galen's system of explanation. Casseri also witnessed in Fabrici the methods of a teacher who frequently "bypassed" actual dissection in order to develop the theoretical category of usefulness. Unlike Fabrici, however, Casseri elaborated on the experience of dissection, that is, on the technical skill it required to produce "so detailed an

FIGURE 13. Giulio Casseri, *De vocis auditus organis historia anatomica* (Ferrara: Victorius Baldinus, 1601), table 1. Reproduced courtesy of the Boston Medical Library in the Francis A. Countway Library of Medicine, Harvard University.

acquaintance with even the smallest parts" of the human and animal bodies that lay on his dissecting table. This empirical knowledge reflects Casseri's work as a surgeon, as well as his participation in the private anatomies that had become a more standard feature of anatomical study in Padua in the late sixteenth century. With more material available for dissection (a history of acquisition that chapter 4 sought to map), private anatomies were not only more frequent than public anatomies, they were also longer and more visually intense studies of anatomy, employing dissection and vivisection.

Casseri continued to give private anatomies. On 3 January 1613–1614, Casseri was given responsibility for the public anatomy, but he chose not to hold it in the public theater; it took place instead "in the room of the Illustrious Prefect, in which a very sumptuous theater in an accommodating space was constructed."[55] This demonstration was supposedly public, but its location in the prefect's room redefines, again, what the term "public" means, for the sumptuous theater in that room was not open to the public in the same way as Fabrici's anatomical theater was. Instead, calling this demonstration "public" signaled Casseri's ascendance up the ladder of academic posts, because he was taking over responsibility for the public demonstration. The demonstration was elaborate, and it was followed by a burial where several orations were given.[56] Nonetheless, the demonstration paralleled the content and pedagogy of Casseri's private anatomies. Casseri worked with student-assistants in the chamber, but they were called *massarii* rather than *anatomistae*, emphasizing their assistance rather than their performance of dissections. Casseri ended his portion of the event with a vivisection of the recurrent laryngeal nerves in a dog.[57] This was a public anatomy demonstration, but it had so many of the features of a private anatomy that its locale might better be called an attenuated private venue: it was sumptuous and decorated, but it focused attention on the demonstrator's display of technical skill.

In *De vocis auditu* (1600 [1601]), Casseri included a chapter on the recurrent laryngeal nerves, and it is tempting to read this chapter as an extension of his canine vivisections. The laryngeal complex, with its several structures, was often illustrated by vivisecting small animals. To explain the movement of the cords by means of the nerves, Casseri compared them to the pulleys of a *glossicomium*. The *glossicomium*, mentioned by Vesalius and others, was a well-known surgical instrument, used, according to

Casseri, by "mechanical laborers as well as the doctors named 'organici'"
to restore fractured and broken bones in the legs and arms:

> The construction of this machine is as follows: its top and bottom sec-
> tions contain a plank, and on the machine several little pulleys are built
> up; to these pulleys is tied in crisscross fashion a series of snoods. Now
> when doctors plan to use it, for example, to reset dislocated bones, they
> encircle the broken limbs with ropes that hand down from the *glossico-*
> *mium* and by a gentle tightening of the ropes a tension in either direction,
> up and down, is effected, which is suitable and correct for the affected
> limbs. In these movements lies what we are calling metaleptic motion.
> Hence it is clear that metaleptic motion consists of raising and lowering,
> whether in elevation or depression, and of contrary, but simultaneous mo-
> tions of the same thing. Our mind hasn't the power even to conceive that
> this motion can be effected without the pulleys.[58]

Casseri concludes the chapter by connecting this example to the position of anatomical structures and their movement, by noting how com-
pressing the nerves produces muteness, and by attending to differences
in substance (relative thickness, moistness). The passage illustrates the
movement of the laryngeal complex by way of an analogy borrowed from
surgery and a device used to set fractures. Casseri seems comfortable
with the idea that surgical instruments such as the *glossicomium* allow
the mind to conceive of relationships between anatomical parts, even
when the parts cannot be viewed on site, performing their functions.

The surgical instrument analogy enabled Casseri to extend the information imparted by the vivisection and the examination of the laryngeal
nerves. Casseri had elsewhere demonstrated surgical operations, but here
he used a surgical instrument to promote a mechanical understanding
of movement, which either could not be seen in the demonstration or
could be seen only fleetingly. Seventeenth-century mechanism was a
multifaceted concept. As Domenico Bertoloni-Meli has explained, anatomists such as Nicolas Steno (1638–1686), in his *Discours*, understood
"mechanical" to mean machinelike attributes rather than movements
based on the laws of mechanics. Therefore, they placed objects at the
center of their work, because objects embodied more abstract relations.[59]
They would then seek to address the complexities of structure by de-
constructing it, handling its parts separately and associating particular

elements with simpler objects. In an attempt to examine a particular motion that could not be stabilized in the process of vivisection, Casseri looked to the mechanical properties of a surgical instrument, the *glossicomium*, a mechanical system of ropes and pulleys that corrected damaged limbs and allowed students to conceive of the metaleptic motion associated with voice.[60]

## CONCLUSION

By the early seventeenth century, the private anatomies at the University of Padua designated an alternative anatomical tradition, though this format suffered from an insufficient supply of corpses, just as the public tradition did. In private anatomies, Casseri's technical expertise was seen not only in his delicate dissections and vivisections, but also in his ability to combine anatomical and surgical instruction. When Casseri demonstrated publicly in the early seventeenth century, his demonstrations retained the pedagogical merits of the private anatomical exercise. In these smaller venues, Casseri turned his audience's attention to the structural details of his specimens, leading them from the whole to the parts. He transformed contemplation, tethering it to the material stuff of anatomy and construing it as a visually oriented exercise. This kind of visual experience was an umbrella covering contemplation, elongated attention spans, and discipline. It also contained the bifurcation between manual acts and passive observations, between Casseri's "applied and operative" method and "the points of usefulness" understood by "mental acuity alone." When Casseri substituted for Fabrici in the public anatomy in 1604, one student noted that Casseri's demonstration was "useful in the most important ways to the students"; "he read to the students and demonstrated ocularly this anatomy . . . everyone was able to see particularly all the parts . . . [and] the ways of treatments." For this he earned "the greatest attention every morning" from his audience.[61] However basic, this comment shows that the students' attention depended on what they could see in Casseri's "ocular" demonstration—what Casseri was able to reveal by virtue of his technical expertise.

At the end of the sixteenth and the beginning of the seventeenth centuries, the increasingly robust nature of private anatomies promoted the students' interest in technical skill and expertise. They celebrated it in their colleagues and sought it out in their professors. William Harvey, who studied in Padua in the late 1590s, began his *Lectures on the Whole of Anatomy* (1616) very conventionally, with a definition of anatomy as "that faculty [that] through inspection and dissection reveals the uses and actions of the parts":

> [There are] five [general] divisions of anatomy: (1) descriptive narrative, (2) use, action and services, whereby there are (3) observations of those things which occur rarely and as a morbid condition, (4) resolution of the problems of authors, (5) skill or dexterity in dissection and the condition of the prepared cadaver.[62]

The first four phases of the inquiry were a standard feature of the growing literature on anatomy in the early modern period. Fabrici, for example, followed them. Harvey's mention of dexterity and skill, as well as judgment about the condition of the prepared cadaver, however, was a comment about things that were familiar to medical students, for they were the ones to note poor or favorable preparations and earnest or slack assistants. Here and elsewhere, Harvey encouraged spectators and students of anatomy to attend to the particulars of a dissected corpse—normal, abnormal, and morbid (or pathological) conditions—but his decision to incorporate technical skill into the definition and divisions of anatomy recalls the anatomical activities of medical students in Padua, where student-assistants were carefully watched by other students and, when it seemed appropriate, praised for their careful preparations, their ability to acquire corpses, and their skill in cutting into the cadavers. In his "canons of general anatomy," Harvey further emphasized the need "to show as much in one observation as can be" and to "cut up as much as may be in the presence [of the spectators] so that skill may illustrate the [descriptive] narrative."[63]

The tradition of private anatomies began to animate the academic community in Padua in the late sixteenth and early seventeenth centuries promoting the careful dissection and vivisection of bodies and a scrutiny of their parts. The compact spaces, the distinct content, and the alternative pedagogy of private anatomies supported the transformation of attendees from silent spectators, such as those earnest transalpine students who stood quietly in the public anatomical theater, to silent observers, such as those, equally earnest, who gazed upon Casseri as he completed the dissection of nine cadavers, the vivisection of a dog, and the "careful" operations of surgery. Students came to Casseri's private anatomies because the events were exclusive and because they rendered the technical expertise associated with anatomy and surgery in both accessible and dramatic terms. Having scoured the city for corpses, preferably fresh ones, students went to the private chambers of Casseri. Once there, safe from the contaminating presence of lowly craftsmen and standing close to the anatomist, they watched him cut, isolate, and manipulate. There, they learned to see what was revealed.

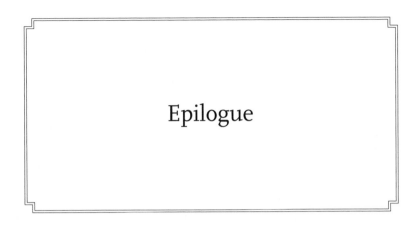

# Epilogue

Neither do Philosophers suffer themselves to be addicted to the slavery of any man's precepts, but that they give credit to their own eyes; nor do they so swear Allegiance to Mistris Antiquity, as openly to leave, or in the sight of all to desert their friend Truth."[1] So writes William Harvey in the dedicatory epistle, addressed to D. Argent, the president of the College of Physicians in London and his "singular Friend," that begins *De motu cordis*. Inevitably, when I ask students to unpack the rhetorical oppositions that structure this passage, they point out that true philosophers are open to ideas rather than glued to preformed ones, and that philosophers are engaged in this world and hew to what they see rather than to the texts of antiquity. These observations are, of course, consistent with the passage and with standard formulations of both the Scientific Revolution and the early modern period. I find myself returning to the opposition between the mistress and the friend, however, wondering why Harvey chose to cast the relationship between antiquity and an implied modernity as an opposition between a female lover and a devoted friend. Mistress Antiquity was not necessarily fickle, but tyrannical, wanting the philosopher to swear allegiance to her and thereby publicly abandon his comrade Truth. Say what you will about Realdo Colombo's discovery of the clitoris, the prostitute's body that was the impetus for his inquiry, and the erotic potential of anatomy; they hardly left a trace in Harvey's letter, in which he frames the pursuit of truth as fraternal rather than erotic. Anatomists and physicians such as Harvey were no longer wedded to the ideas of ancients, but instead to their friend Truth, even if true knowledge was splintered and partial: "nor are they [true philosophers] so narrow spirited to believe that ever any art or science was so absolutely and perfectly taught in all points, that there is nothing remaining to the industry and diligence of others."[2]

Harvey's dedicatory letter transformed the erotic charge and the singular obedience that Mistress Antiquity failed to muster into a desire for truth, shared among friends: by his friend D. Argent and by his fellow physicians. In the context of Harvey's ocular demonstrations, the patrilineal model for the production and transmission of knowledge no longer adhered to a strict hierarchy. Harvey was the writer and the producer, but, throughout the text, he cast the demonstrations' spectators as witnesses and as coproducers of anatomical knowledge. The patrilineal model of transmission has at least two basic versions, both hierarchical and both captured by Vesalius in his *Fabrica*: the first, where Vesalius steals the truth from the female body (such as the one exposed to view on the title page of the *Fabrica*; see figure 4 in chapter 1); and the second, where he transmits that truth to his followers, as if from father to son (a relationship and mode of transmission made explicit in the dedication and rendered figuratively in the title page).[3] While the university and the college of physicians were very different kinds of institutions, and while the former may well have been more hierarchical than the latter, anatomical culture in late sixteenth-century Padua encouraged medical students to collaborate in the procedures for acquiring corpses and in the processes for dissecting them. This culture taught students how to behave, that is, how to comport themselves inside anatomical theaters and at the edge of the dissecting table (where increasingly they became witnesses and coproducers of anatomical preparations and knowledge). It also fostered an appreciation of technical skill and expertise in dissection, but it located the sources of expertise in advanced and especially gifted medical students, as well as in professors.

Scholars, in highlighting Aristotelian studies of nature and technologies associated with the pump and with locks, continue to wonder about the influence of Padua and its university on William Harvey.[4] I would like to suggest that Harvey's approach to the scenes of anatomy and to his spectators echoes some of the main developments of the post-Vesalian era of anatomical inquiry. Over the course of this book, I have sought to defend the notion that the anatomists working in Padua after Vesalius's departure were not in any explicit way following Vesalius; they, too, like Vesalius, were trained in Galenic medicine and focused on a Galenic study of anatomy. Secondly, I have attempted to show that the changes in anatomical study between 1550 and 1620 did not depend on increasingly elaborate endorsements of Vesalius's anatomy. They relied instead on texts, teaching practices, and the combination of new and old instructional spaces. Thus the period between 1550 and 1620 is no more Vesalian or post-Vesalian than it is Fabrician. Rather, this period witnessed unprecedented institutional support for anatomy—to maintain its reputation abroad and to satisfy its students. As the support for anatomical exercises grew and professors such as Fabrici competed for resources and recognition, a culture developed within the academy (and especially among students) around the acquisition of corpses and the preparation of cadavers that made it possible to acquire and dissect more cadavers than ever before. The public anatomy demonstration was ill situated to promote or handle this bounty, and a tradition of private anatomies was gradually formed. Even when anatomists demonstrated in a quasi-public venue, such as Casseri in the early seventeenth century, the content of the demonstration was derived from their private inquiries. The changes in anatomical study reflect the different uses of dissection—to study general anatomy, to focus on topics and groups of structures, to practice the techniques of dissection and then vivisection, and to connect anatomy to surgical operations. The medical students' records do not indicate great interest in pathological discussions linked to anatomy, though this could be a feature of the documents, which focus on the stages of medical training before graduation and before apprenticeships in medical colleges.

In this book, I have explored the history of anatomy as a series of intersections between professors, students, printed anatomical texts, and accounts of medical practice. While I have focused on the academic tradition, I have also characterized the academic study of anatomy as dynamic and evolving, constantly influenced by administrative and civic decisions and frequently porous to its surrounding culture. The records of medical students draw out that dynamism and sometimes (hopefully) provide comic relief. Perhaps above all, I am committed to an account of the anatomical theater that neither depends on our own distance from the dead nor bends too quickly to the simultaneous fascination and fear that we seem to feel for a dissected corpse. Instead, by highlighting the embedded nature of early modern spaces, I have reconstructed several perspectives on the anatomical theater, the collective experiences that students shared inside and outside the theater, and the meaningful changes that the theater brought about in the study of anatomy, surgery, and medicine in the early modern period.

# Acknowledgments

This book is not an updated version of my dissertation. Instead, it is a new book for me and for the field about how the anatomical theater served medical students and how medical students learned to dissect bodies and study anatomy long after Vesalius was dead. This work began in Berlin during a postdoctoral fellowship at the Max Planck Institute for the History of Science (MPI). Its early conception was influenced by the director of the MPI, Lorraine Daston, and by the participants in the Common Languages of Art and Science research group. My earlier studies in graduate school at the University of Chicago prepared me to undertake this project. I wish to acknowledge and to thank my advisors, Robert Richards and Michael Murrin, who kindly but firmly guided my inquiries into the histories of science, cultural history, the history of the book, and Renaissance literature. Elissa Weaver provided a consistently challenging and rigorous model for combining historical and literary inquiry. Noel Swerdlow was a source of enthusiasm (even when my own waned) and encouragement as well as erudition; I am grateful for his continued support.

Despite their neat edges, books signify complex processes and long stages of research, thinking, and writing, followed by more research and constant critiques. Since my postdoc, I have benefited not only from the provocative and learned studies of Katharine Park, but also from conversations and correspondence with her. She has animated central and marginal features of this project and shaped its development in countless ways, for which I am especially grateful. In addition, I wish to thank the anonymous reader for the Johns Hopkins University Press for comments on each of the chapters, which motivated me to strengthen the arguments therein. A number of scholars have discussed key concepts and graciously offered feedback on different chapters of this book. I thank

Andrea Carlino, Andrew Cunningham, Jonathan Davies, Mary Fissell, John Krige, Rafael Mandressi, Gideon Manning, Eileen Reeves, and Massimo Rinaldi. At Miami University, I have the privilege of participating in the Early Modern Collective and benefiting from the critical, insightful eyes of its members. Audiences at the University of California at San Diego, the California Institute of Technology, Frei University, the University of Paris–Lyon, and the University of Sydney have helped to shape further various sections of this book. Very special thanks are due to Stephen Nimis, professor of Classics, and Edgar Evan Hayes, rare marvel of an undergraduate student of Classics, whose Latin expertise not only saved me from many infelicities and errors but prompted further, useful reflection on the accounts of medical students. I would also like to thank my editor, Jacqueline Wehmueller, for her targeted insight, my copyeditor, Kathleen Capels, for her diligence and encouragement, and the editorial staff at the Johns Hopkins University Press for their expertise.

One of the delights of academic work, which is also the source of great intellectual and emotional support, is the community of friends and colleagues that I have been fortunate enough to be a part of over the years. They are connected to different places (such as Berlin, Venice, Atlanta, Oxford-Ohio, and Florence), different times, and different institutions, but each has made the long travail not only instructive, but pleasing as well: Nadja Aksamiji, Andreas Corcoran, Elisabetta Cunsolo, Una and Tony Delia, Filippo de Vivo, Claire Goldstein, Kimberly Hamlin (the perfect writing partner), Narin Hassan, Elizabeth Horodowich, Frederick Ilchman, Ken Knoespel, Ruth Mack, Gerry Milligan, Peter Mitchell, Joshua Phillips, Meredith Ray, Mary Quinn, Martha Schoolman, Bridgette Sheridan, Jonathan Simon, Sharon Strocchia, Bette Talvacchia, Colleen Terrell, Nick Wilding, Carrie Yury, and Emily Zelner.

In addition, this project has received the support of several institutions: the Delmas Foundation for research in Venice; the American Association of University Women; the Max Planck Institute for the History of Science; the Huntington Library; the National Endowment for the Humanities; the Renaissance Society of America; the American Council of Learned Societies; and Harvard University for research fellowships at the Countway Library and at the Villa i Tatti. I thank these institutions and the several individuals at American and Italian libraries and archives who helped me find sources and decipher them: especially Emilia

Veronese and Maurizio Rippa Bonati in Padua, Jack Eckert in Boston, and Anna Smith in London.

This project would not have taken off without the support of my family. My greatest thanks go to Andrew Hebard, whose capacity for insight remains breathtaking.

# Notes

PREFACE

1. Trivedi V. N. Persaud characterized Vesalius's successors as dependent on Vesalius and responsive to his innovation; in this sense, Vesalius's successors have been understood as post-Vesalian. See Persaud, *History of Anatomy*. While bibliographical details on Vesalius will be given in chapter 1, the standard biography remains O'Malley, *Andreas Vesalius of Brussels*.

2. See the introduction to *Embryological Treatises*. On Fabrici's interest in natural philosophy, see Cunningham, "Fabricius and the 'Aristotle Project.'" On Fabrici as a teacher, see G. Favaro, "L'insegnamento anatomico."

3. The descriptions in travel literature are discussed in chapter 3. On the role of anatomy in Padua in the sixteenth century, see Bylebyl, "School of Padua."

4. This line of thought, now tarnished and withered, was derived from positivist accounts of the Scientific Revolution. Vesalius is still called the first modern anatomist, albeit as a slight qualification of his earlier designation as the "father" of modern medicine. For example, see Osler, *Evolution of Modern Medicine*.

5. On Vesalius's humanism, see Carlino, "Les fondements humanistes." For visuality in relation to Vesalius, see Sawday, *Body Emblazoned*, 16–38; and Park, *Secrets of Women*, 207–60.

6. Science and civility have been studied in the context of early scientific communities, such as the Royal Society in England. These communities, it has been argued, were decoupled from their nearby universities and were just distant enough from the practices of disputation, source criticism, and contemplation (all of which were integral to academic life) that they generated new forms of inquiry. Instead, my research shows that these practices, while very much a part of the university, also allowed new affects and styles of inquiry to emerge within the university. On the Royal Society, see Shapin, *Social History of Truth*. On the critical genealogy of civility, see Burckhardt, *Civilization of the Renaissance*; and Elias, *Civilizing Process*.

7. Daston, "Moral Economy of Science"; and Daston, "Baconian Facts."

8. In addition to his bringing a range of new documents—and even a whole archive—to light, Richard Palmer has underscored the learned dimensions of surgery as it was practiced and studied in the medical colleges in Venice. See Palmer, *Studio of Venice*. In

chapter 5, I attempt to highlight the professional concerns of medical students as they relate to the academic study of anatomy and surgery.

9. In *Studio of Venice*, Palmer has shown that medical students began to take their degrees through the *studio* in Venice, which cast the *studio* as an institutional rival to the university. The activities of anatomists in Padua and a widespread interest in private anatomies suggest that this university attempted to respond to the success of the *studio* in Venice.

## INTRODUCTION

1. The original letter is in *Lettere dei Riformatori dello studio*, held in the Archivio di Stato, Venice (hereafter cited as ASV). The ellipsis indicates a corrupted text: "Gli tempi serenissimi aggiuntavi la neve, et il . . . chio, colle vacanze di Natale che vengono invitano all'anatomia, et dopiano l'ardore immenso de scolari, gli quali essendo stato due anni senza, non veggono l'hora che si venga a mostrargli la fabrica humana: Qua sono molti scolari Thedeschi et Poloni, gli quali non vedendo preparamento alcuno, et dubitando quasi che non si faccia per diffetto di soggietto [missing cadavers], s'incominciano d'apparecchiare per andare a Bologna o a Ferrara, dove indubitatamente l'havranno queste feste. Io gli vado tratenendo con buone promesse, et che senza fallo in queste vacanze havremmo l'anatomia; ma non so poi come attendergli se non sono aiutato dalle Ill. Me Mag V." Falloppio's letter was published in G. Favaro, *Gabrielle Falloppia modenese*, and again in *Epistolario di Gabriele Falloppia*.

2. Park, *Secrets of Women*.

3. With regard to the history of anatomy, older historiographical approaches tend to focus primarily on published texts, especially those of Vesalius and Harvey, emphasizing Vesalius's celebration of dissection and its culmination in Harvey's discovery of blood circulation. Accordingly, the revolutionary part of the Scientific Revolution designated Vesalius's aggressive critique of ancient authorities and his methodological aspects as being most evocative of modern scientific practice. Not only do these approaches (which depend heavily on the history of ideas) elide the shared intellectual traditions of the period, but they also tend to view the role of teaching as perfunctory and the role institutional inertia played in the curriculum as secondary (or adversive) to the *real* science at hand. Until recently, the conditions in the classroom have not been seen as instruments of historical change. With my focus on teaching and on the responses that certain lessons elicited from students, I aim to rectify that critical gap in this book. On historiographical modes in the history of science, see Cohen, *Revolution in Science*, 176–96; Findlen, "Between Carnival and Lent"; Dear, "Towards a Genealogy"; and Harkness, *Jewel House*, 1–14.

4. See Daston, "Attention."

5. *Atti della nazione germanica.*

6. Sawday, *Body Emblazoned*, 1–15, 54–84.

7. Richard Sugg notes that some books, such as Helkiah Crooke's *Microcosmographia*, "both reflected and stimulated an immediately visual relationship to the anatomized body." See Sugg, *Murder after Death*, 1, 4.

8. Camporesi, *Anatomy of the Senses*, 130–46; and Henderson, *Renaissance Hospital*.

9. Mandressi, *Le regard de l'anatomiste*, 83–92.

10. Findlen, "Sites of Anatomy," esp. 274–80.

11. On anatomical illustration, see Choulant, *History and Bibliography*; Lambert, Wiegand, and Ivins, *Three Vesalian Essays*; Muraro, "Tiziano"; Kemp, "Mark of Truth"; and Kemp, "Temples of the Body."

12. For a discussion of anatomy and therapeutics, see Bylebyl, "School of Padua." On the sacred origins of dissection, see Park, *Secrets of Women*, 39–76. On the relationship between anatomy and cartography, see Sawday, *Body Emblazoned*, 16–38; and Traub, "Mapping the Global Body."

13. Following Bakhtin's *Rabelais and His World* and Foucault's *Discipline and Punish*, Sawday provides an account of anatomy in relation to penal codes and infamy. See Sawday, *Body Emblazoned*, 54–84. On the overlap of anatomy and Carnival, see Ferrari, "Public Anatomy Lessons." On Pisan practices and the relationship between anatomy and burial, see Lazzerini, "Le radici folkloriche."

14. Katharine Park underscores the origins of the taboo against the disruption of burial in "Criminal and the Saintly" and in "Life of the Corpse." Andrea Carlino explores this taboo as being anthropological and notes two ways that institutions sought to mediate it: first, by controlling the processes for obtaining cadavers, and second, by making the new practices continuous with the older practices associated with autopsy. See Carlino, *Books of the Body*, 92–114.

15. See n. 13.

16. Ferrari, "Public Anatomy Lessons," 52.

17. Ibid., 74–82.

18. Sawday treats London's anatomies in *Body Emblazoned*, and Leiden's in "Leiden Anatomy Theatre." Ferrari's "Public Anatomy Lessons" treats the Bologna tradition; Lazzarini's "Le radici folkloriche," the Pisa tradition; and Carlino's *Books of the Body*, the tradition in Rome. On Padua, see chapter 3. For a study of the anatomical theater in the Spanish context, positioned between the university, professional life, and government, see Martínez-Vidal and Pardo-Tomás, "Anatomical Theatres."

19. See Rupp, "Matters of Life."

20. Saibante, Vivarini, and Voghera, "Gli studenti dell'università"; A. Favaro, *Galileo Galilei*; and Kagan, "Universities in Italy."

21. The decree is found in *Padova, Studio. Documenti 1467–1625*, held in the Library of the Correr Museum, Venice: "Ricerca la dignità è servitio del studio nostro di padoa che li dottori forestieri condotti da queste cons. per erudir li scolari in queste scienze che tal sono necessarie per ben regere ogni Republica, essendo queste persone conspicue di gran valore, è stima appreso il Mondo tutto, sino honorati et ben trattati da ogni anno et spetialmente da gli altri dottori della Citta."

22. David Kaiser emphasizes pedagogy as a way to move between generalist and microhistorical approaches to the history of science. See Kaiser, "Focus," esp. 250; and his introduction to *Pedagogy*, 1–11. See also Becher, *Academic Tribes and Territories*.

23. *Pedagogy*, 1–11. For an accessible introduction to the Scientific Revolution that begins with a university education, see Dear, *Revolutionizing the Sciences*.

24. These students came from Padua, the Italian peninsula, and all over Europe. In his research, Biagio Brugi includes Padua, Rome, Lucca, Calabria, Abruzzo, Sicily, Lombardy (including Milan), Cremoso, Genoa, Piacenza, Tuscany, Piemonte, Savoy, Saluzza, Monferrato, Trevisa, Friuli, Istra, and Venice; and for transalpine regions, the German-Silesian regions, Sweden, Russia, Bohemia, England, Scotland, and France. See Brugi, *Gli scolari dello Studio*, 24–25.

25. On style, see M. Shapiro, "Style"; Crombie, *Styles of Scientific Thinking*; and Hacking, "'Style' for Historians." For anatomy, see Klestinec, "History of Anatomy Theaters."

26. See Agrimi and Crisciani, *Edocere medicos*.

27. Vesalius, *De humani corporis fabrica*, dedication, 2v: "His quidem graculorum modo, quae nunquam aggressi sunt, sed tantum ex aliorum libris memoriae commendant, descript ve ob oculos ponunt, alte in cathedra egregio fastu occinentibus."

28. Ibid.: "Gotthorum illuviem . . . medicina eo usq lacerari cepit, quod primarium eius instrumentum manus operam in curando adhibens, sic neglectum est, ut ad plebeios ac disciplinis medicae arti subservientibus neutiquam instructos, id quasi videatur esse demandatum."

29. Fabrici, *De venarum ostiolis*, 46: "Eedetur, sibi compararint suisque locis collocarint ac disposuerint, tandem universos in unum volumen coniectos apte compingere et colligare sine ulla aut libri, aut impense, iactura."

30. On Fabrici's *theatrum* (without reference to theaters of memory or dramatic theaters constructed during the period), see the recent exhibition catalog, *Il teatro dei corpi*. On the organic soul, the life force responsible for the body's processes of digestion, sensation, movement, respiration, and generation, see Park, "Organic Soul." On theaters of memory, for which there is a growing bibliography, see Yates, *Art of Memory*; Carruthers, *Book of Memory*, 56–98, esp. 90; and West, *Theatres and Encyclopedias*.

31. Many historians have noted this feature in the history of anatomy. On its early development, see Siraisi, *Taddeo Alderotti*; and on the sixteenth century, see Carlino, *Books of the Body*, 120–28. For a detailed overview of Padua, see Ongaro, "La medicina." The study of a range of classical and medieval works by philosophers and medical writers, including Hippocrates, Aristotle, Galen, and Avicenna, was important to medical theory. Paul Grendler provides a sample of the curriculum in Bologna, and also the curriculum in natural philosophy, in Grendler, *Universities*, 320–21, 269–79.

32. Vesalius, *De humani corporis fabrica*, dedication: "Caeterum perversissima haec curationis instrumentorum ad varios artifices diductio, adhuc multo execrabilius naufragium, ac longe atrociorem cladem praecipuae naturalis philosophiae parti intulit (2v) . . . Ac ne, omnibus aliquid communium studiorum gratia tanto successu attentantibus, solus torpescerem, aut etiam a meis progenitoribus degenerarem, hoc naturalis philosophiae membrum ita ab inferis revocandum putavi, ut si non absolutius apud nos, quam alias unquam apud priscos dissectionum professores versaretur (3r)."

33. On theoretical and practical medicine, see Siraisi, *Medieval and Early Renaissance Medicine*, 48–77.

34. Grendler, *Universities*, 322. On Avicenna's *Canon*, see Siraisi, *Avicenna*, 54.

35. See *Embryological Treatises*, 192.

36. Fabrici, *De visione, voce, auditu*, "De aure," proem.

37. The public-private distinction is often noted, but it has also generated error. Jonathan Sawday, for example, in *Body Emblazoned*, assumes that public anatomies proceeded as public autopsies, an idea that conflicts with the academic tradition of using public anatomy demonstrations to introduce students to the general features of human anatomy (and not the particular features that caused the death of the cadaver).

CHAPTER 1: SPECTACULAR ANATOMIES

1. Gabriele Falloppio took the chair in surgery, simples, and anatomy in 1551. For more on Falloppio, the standard biography remains G. Favaro, *Gabrielle Falloppia modenese*.

2. See Introduction, n. 1.

3. G. Favaro, *Gabrielle Falloppia modenese*, 227: "Però prego quelle riverentemente, che voglino scrivere una sua al clar.mo Podestà et raccommandargli l'anatomia chiedendogli un soggietto quanto più presto sarà possibile, o con . . . a gli Massari dell'Anatomia, che essi segretamente se ne passano procacciare uno quando gli venga l'occasione di person ignobile et non conosciuta." On 7 December 1556, Falloppio sent a letter to the rectors of Padua (civic officials), recommending that anatomical specimens be procured. Then, on 15 December 1556, the *riformatori* (liaisons between the university and Venice) wrote to the *podestà* (another civic official) to request anatomical specimens, and approximately two weeks later the specimens were obtained.

4. The term comes from Persaud's *History of Anatomy*, where it serves to organize the history of anatomy before and after Vesalius according to the ways that anatomists prefigured and then continued Vesalius's anatomical work.

5. G. Favaro, *Gabrielle Falloppia modenese*, 227: "Io prometto alle Illme MV di fare con gl'orsi et la simia una bellissima Anatomia in questi 30 giorni che non si leggerà, et compirla prima che vengano le lettioni dell'anno nuovo."

6. Ibid.: "Et havrei caro, che quelle fossero presenti alla fattica et diligenza che userò in mostrare questi occulti misteri d'Iddio."

7. "The secrets of nature" was an old metaphor and a consistent part of the anatomical tradition. For an account of how the secrets were generated, maintained, and adapted in the medieval traditions of anatomy, see Park, *Secrets of Women*. Related to this is the way that the secrets tradition generated a metaphor for the acquisition of knowledge, as a hunt (*venatio*). Falloppio's emphasis on the work of discovering hidden secrets bears the signs of this tradition as well. See Eamon, *Science*, 269–300.

8. On the technical arts, see Long, *Openness, Secrecy, Authorship*. On the trades and professions of early modern Venice and Italy, see Pullan, *Rich and Poor*; and McClure, *Culture of Profession*, esp. 142–77.

9. *Statuta almae universitatis*, bk. 2, ch. 27, 36. Also see Bylebyl, "Interpreting the *Fasciculo*"; Nardi and Musatti, "Dell'anatomia in Venezia," 6; and Tosoni, *Della anatomia degli antichi*, 103–4.

10. The position of university rector was filled by a student elected by his fellow students, but the anatomist was chosen by rotation among the professors of the four universities. The student rectorship was later taken out of the hands of the students, usually when a university came under foreign governance. This position died out after a series of

educational reforms initiated by Maria Theresa in 1771. For a discussion of the position, see Ridder-Symoens, "Management and Resources," 171–72.

11. *Statuta almae universitatis*, bk. 2, ch. 27, 36. In Venice, the Senate confirmed the practice of anatomical study in the statutes of the medical college in 1507. This confirmation suggests the existence of a tradition of demonstration, one superior to the exercises of anatomy in Venice as "secundum legum et consuetudinem antiquam." Jerome Bylebyl notes that "the Venetian statute dates back at least to 1368, when the Venetian government ordered the College of Surgeons to conduct an anatomy at least once a year, which all physicians and surgeons of the city were required to attend, and to 1370, when it ordered the physicians to share equally in the expenses." See Bylebyl, "Interpreting the *Fasciculo*," 311. For a discussion of the organization as it is reflected in the iconography of anatomy scenes, see Carlino, *Books of the Body*, 8–68.

12. Ridder-Symoens, "Management and Resources," 168.

13. On the student nations in Bologna and Padua, see Kibre, *Nations*, 3–64, 116–22.

14. On the licit means of acquiring cadavers, see Carlino, *Books of the Body*, 77–91. On the illicit means by which cadavers were obtained, see Park, "Criminal and the Saintly."

15. *Statuta almae universitatis*, 36. For a full discussion of these, see Bylebyl, "Interpreting the *Fasciculo*," 309–11.

16. According to Andrea Carlino, the dissector could have been a barber, a surgeon, a graduate surgeon, or a student. On the format, see Carlino, *Books of the Body*, 11–12.

17. *Statuta almae universitatis*, 36.

18. Andrea Carlino characterizes the inertia of the textual tradition of anatomy throughout his *Books of the Body*. Most recently, Rafael Mandressi has traced the confrontation between the book and the anatomist as sources of authority; his analysis reconstructs the gaze of the anatomist as a mediating device for this confrontation. See Mandressi, *Le regard de l'anatomiste*.

19. Mondino, *Anothomia*; and Ketham, *Fasciculo di Medicina*.

20. Carlino, *Books of the Body*, 91.

21. Berengario da Carpi, *Commentaria*, 119v, 479v, 516r. On Berengario da Carpi himself, see French, *Dissection and Vivisection*, 96–98; and Park, *Secrets of Women*, chapter 1.

22. Berengario da Carpi, *Commentaria*: "quia certe anatomia communis non monstrat nobis perfecte omnia requisita cuilibet membro: et sic ego feci pluries et optime cognovi quae cognoscenda erant de eis membris: veritas est que diligens et expertus operator multotiens bene cognoscit quae non cognoscerentur sine tali diligentia in anatomia communi" (119v); "maxime in demonstrando anatomiam secundum que fit anatomia in gimnasiis publicis: quia non potest demonstrari in tali anatomia numerus musculorum: neq ossium: neq nervorum: neq venarum: sed ut inquit Mundinus hic et alibi talia membra melius videntur in corporibus eliquatis in aqua fluminis: vel in copore decocto aut exsiccato quam in anatomia communi ab eo tradita" (479v); "Nota lector que eo fui diminutus in anatomia musculorum totius corporis: quamvis de aliquibus particularibus musculis completum fecerim sermonem: et non multum me intricavi in anatomia omnium musculorum: quia in ostensione communi anatomica quae fit in gymnasiis pro scolaribus maior pars musculorum non potest monstrari: sed requititur ad eam perfecte videndam extremus labor: et tempus longum: et ita locus accomodatus" (516v).

23. Baldini, "Per la biografia."

24. Because private lessons were extracurricular and often took place in venues beyond the university campus, they have some features in common with later learned societies, which were not directly linked to university education. For the English case, see Shapin and Schaffer, *Leviathan and the Air-Pump*; and Shapin, *Social History of Truth*. For the French, see Stroup, *Company of Scientists*. For the Italian, see Freedberg, *Eye of the Lynx*.

25. On the role of autopsy in the traditions of medieval piety and sanctity, see Park, *Secrets of Women*, 39–76. For treatment of a later period, see Pomata, "Malpighi."

26. Sawday, *Body Emblazoned*, 6–15.

27. See the discussion of Matteo Corti later on in this chapter, as well as Heseler, *Andreas Vesalius' First Public Anatomy*.

28. Heseler, *Andreas Vesalius' First Public Anatomy*.

29. Cunningham, *Anatomical Renaissance*.

30. See M. Shapiro, "Style."

31. Heckscher, *Rembrandt's Anatomy*, 46: "The all-over stylization of the show anatomies was also reflected in their outward forms, in the careful distribution of roles among those engaged in the dissecting, demonstrating, reciting, and disputing, and in the elaborate protocol that ruled over the seating of those in attendance—the 'distributio,' to use the term employed as early as the fifteenth century, 'sedendi ordinis pro dignitate.'" On the charisma of the professor as a result of institutional changes in the late 1800s, see Clark, *Academic Charisma*, 76–92. For several examples of early modern charismatic professors (and a review of Clark's monograph), see Haugen, "Academic Charisma," esp. 204–9.

32. See Mortimer, "Author's Image," esp. 53–62.

33. Andrea Carlino argues for this reading of Ketham's scene by analyzing the variations on this image in the early editions of Ketham's work. See Carlino, *Books of the Body*, 8–18. Based on an interpretation of the statutes and of the youthful appearance of the chaired figure and the aged one of the figure doing the dissecting, Jerome Bylebyl argues that the positions or roles were more fluid. See Bylebyl, "Interpreting the *Fasciculo*."

34. On medical humanism, see *Medical Renaissance*; Bylebyl, "School of Padua"; Bylebyl, "Medicine, Philosophy, and Humanism"; Nutton, "Rise of Medical Humanism"; and, most recently, Siraisi, *History, Medicine, and the Traditions*.

35. For a survey of early medicine and its standard curriculum, see Siraisi, *Medieval and Early Renaissance Medicine*, 55–77. Paul Grendler provides a sample of the curriculum in Bologna, and also the curriculum in natural philosophy, in *Universities*, 320–21, 269–79.

36. Benedetti, *Historia corporis humani*, ch. 1: "Loco praeterea amplo, perstatili, temporarium theatrum constituendum est, circumcavatis sedilibus, quale Romae ac Veronae cernitur, tantae magnitudinis, ut spectantium numero satisfaciat: ne vulnerum magistri, qui resectores sunt, a multitudine perturbent. Hi solertes esse debent, quiq; saepius resecaverint. Sedendi ordo pro dignitate distribuendus est." The text is available in Italian (Benedetti, *Anatomice*) and in English (Lind, *Pre-Vesalian Anatomy*). On Benedetti

himself, see Ferrari, *L'esperienza del passato*. The translation is from Lind, *Pre-Vesalian Anatomy*, 83.

37. Baldassar Heseler described the setting as a classroom or auditorium where the rector was typically elected (the election itself was a formal assembly with a specified protocol). Heseler, *Andreas Vesalius' First Public Anatomy*, 85: "A table on which the subject was laid, was conveniently and well installed with four steps of benches in a circle, so that nearly 200 persons could see the anatomy"; and no one "was allowed to enter before the anatomists, and after them, those who had paid 20 sol. More than 150 students were present and D. Curtius, Erigius, and many other doctors, followers of Curtius. At last, D. Andreas Vesalius arrived, and many candles were lighted, so that we all should see."

38. Benedetti, *Historia corporis humani*, ch. 1: "mediocris aetatis, non gracilis, non obesi corporis, staturae maioris, ut huberior materia, evidentiorq; sit spectantibus." The translation is from Lind, *Pre-Vesalian Anatomy*, 83.

39. Vesalius, *De humani corporis fabrica*, 548. The translation is from O'Malley, *Andreas Vesalius of Brussels*, 343.

40. Ibid.

41. Andrew Cunningham argues that the sense of theater in Benedetti's work pertains to seeing, but in this chapter I suggest not only that sight was repeatedly impeded by the crowd, but also that the concept of theater pertained to Benedetti's performance and the professional concerns it sought to mediate. See Cunningham, *Anatomical Renaissance*, 167–90. A more recent work is Mandressi, *Le regard de l'anatomiste*, 7–18.

42. Benedetti, *Anatomice*, bk. 3, ch. 1: "Huic laeta felicitas originem ducit, huic mortis terricula abigunt, atq; ad divina mens sublimis penetrat arcana. Ad has partes hac nocte in hoc ipso theatro presentes esse volo senatores meos sapientes ex patritia gente Veneta, Bernardum Bembum, Antonium Boldum equites; Antonium item Calvum triumvir, Petrum Priolum senatorem, qui mecum divinam cordis officinam contemplentur, et arcana naturae perquirant, iam perspectis et compositis reipublicae rebus, et domum fessi, sed hilares abeant."

43. Massa, *Liber introductorius anathomiae*, 6: "Quod si deus mihi (ut fecit) sua misericordia auxiliabitur, reliquas corporis partes iu alio copioso scribam volumine. Neq; expectare te volo a me scamni modum praeparationis, aut sedilium, ubi spectantium turba te secantem, aut demonstrantem intueatur, seu talia ludibriosa, ne potius me rudem, et non philosophum omnibus meis scriptis ostendam."

44. Andrea Carlino has called attention to this tradition of skepticism toward medicine, noting that although Petrarch's version is the most famous example, this trend exists as a countertradition that runs uninterrupted from the classical to the modern world and contains coherent arguments, topoi, and exempla. See Carlino, "Petrarch."

45. Mandressi, *Le regard de l'anatomiste*.

46. Palmer, "Nicolò Massa," esp. 394–95. On the *Liber introductorius anathomiae*, see French, *Dissection and Vivisection*, 132–37. On the college of physicians, see Palmer, "Physicians and Surgeons"; and Palmer, *Studio of Venice*.

47. Palmer, "Nicolò Massa," 395.

48. Palmer, "Physicians and Surgeons," 458.

49. William Clark discusses the intimacy of the private settings for lectures, informal disputations, and teaching during the early modern period, giving special attention to German institutions. See Clark, *Academic Charisma*, 143–58.

50. In 1537, Vesalius came to Padua and, in the same year, was awarded his doctorate in medicine. Though there is no official record of his appointment, he seems quickly to have been made demonstrator in anatomy and lecturer in surgery, roles he held until 1542, when he went to the printer Oporinus in Basel to oversee the publication of his work, *De humani corporis fabrica*, and the accompanying atlas and abbreviated text known as the *Epitome*. For a biographical study, see O'Malley, *Andreas Vesalius of Brussels*. On Vesalius's humanism, see Carlino, *Books of the Body*, 39–68; and Cunningham, *Anatomical Renaissance*, 88–142.

51. Vesalius, *De humani corporis fabrica*, ch. 19, 547–48. The translation is from O'Malley, *Andreas Vesalius of Brussels*, 342–43.

52. The details regarding this woman are given at the end of chapter 5 of *De humani corporis fabrica*, where Vesalius discusses the uterus. In addition, see Park, *Secrets of Women*, ch. 5.

53. Galen, *On anatomical procedures*; and Galen, *On the Usefulness of the Parts*, 119.

54. Vesalius, *De humani corporis fabrica*, dedication, 4r. I have followed the English translation in Vesalius, *On the Fabric of the Human Body*, vols. 1–3.

55. It is not known whether Vesalius read Massa's *Liber introductorius anathomiae*.

56. Vesalius, *De humani corporis fabrica*, dedication.

57. Ibid., bk. 1, ch.1 ("On the Function and Differentiation of Bones"), 1–3: "Porro satius erit omnes differentia, quas a forma petere integrum esset (cum innumerae occurrant) ad privatas ossium descriptiones refervare. Arduum quippe esset, ossibus nondum enarratis assequi, quae nam horum sint aspera: ut ea quae lapidea in calvariae basi vocabimus, quod praeruptae rupi similia videantur. item quae laevia sint, ut verticis ossa, frontis os, pectoris os. deinde quae triangulum referant, ut scapulae: et quadrangulum, ut verticis ossa: et quae cunei speciem obtineant, ut capitis os, a cuneo . . . et quae iugis assimilentur . . . iugalia autem nostris appellata: et quae S nostrum imitentur, ut claviculae: et quae ensis effigiem ostendant, ut pectoris os . . . et quae radii quo latiores cordulae texuntur figuram exprimant, ut cubitii os radii nomine donatum: et quae cubo tesseraeq comparamus, ut pedis os a cubi imagine . . . et quae molae, scuto et patellae similia dicantur, ut os genu articulo praepositum: ad haec quae totius Italiae circunscriptionem leviter proponant, ut femur: et quae fibulam repraesentent, ut tenuius in tibia os fibula appellatum: et quae coccygis seu cuculi avis rostro comparentur, ut sacro ossi suppositum os, quod coccyx nuncupatur . . . eiusque generis ossium quae forma invicem varient permulta, non ita obiter ab illo cui ossa adhuc incognita sint, intelligerentur."

58. He also could have emphasized anomalies. More frequently than the first, the second edition of *De humani corporis fabrica* takes up particular anatomical structures that are anomalous. See Nancy Siraisi, "Vesalius and Human Diversity." For a contrast, see Colombo, *De re anatomica*.

59. Vesalius's invitation was both an attempt to make Bologna's university compete with Padua's and a sign of Vesalius's growing fame. This point is noted in Heseler, *Andreas Vesalius' First Public Anatomy*, 37.

60. Heseler, *Andreas Vesalius' First Public Anatomy*, 54: "Vide ergo quam fuerit cupidus carpendi et se in rebus eciam nihili et friuolis ostentandi, vanam quarens semper gloriam."

61. Ibid., 290–91: "potestis sic ex nuda demonstratione discere, nisi vos ipsi in manus vestras subiecta acceperitis, etc."

62. Ibid., 290–93: "Ego vidi quomodo cor canis in altum saliebat, et cum non amplius moveretur moriebatur illico canis." Andrew Cunningham cites the event in order to discuss Vesalius's interest in vivisection in relation to Galen. See Cunningham, *Anatomical Renaissance*, 114–15.

63. Heseler, *Andreas Vesalius' First Public Anatomy*, 292–93: "Isti furiosi Itali hinc inde trahebant canem quod nemo poterat vere tangere istos duos motus. Interrogabant autem quidam scolares Vuesalius, quenam sit veritas horum motuum, quodnam ipse crederet, an arterie sequerentur motum cordis, vel diversum cum eo habeant. Respondit Vuesalius: ego nolo hoc proferre, tangatis vos ipsi vestris manibus, et his credite. Tam invidum semper eum esse dicebant, quod tamen Itali non faciebant, sed omnia quae sciebant ostentantes se omnibus publice dicebant."

64. Ibid.

65. On the opposition between observation and book learning, see Mandressi, *Le regard de l'anatomiste*, 61–76.

66. Heseler, *Andreas Vesalius' First Public Anatomy*, 292–93.

67. On humanism, rhetoric, and Vesalius, see Carlino, "Les fondements humanistes."

68. Heseler, *Andreas Vesalius' First Public Anatomy*, 31 n. 1: "[31 July 1540]: Quod si hodie sunt, qui nil nisi quod factum, atque illecebris lingue, lectorem demulceat, in scholis recipiendum esse ducant, illis per me liberum erit, Alexandri Benedicti latinissimos de resectione libros, Io Andernaci atque Andreae a Lacuna, item Vesaliensis et reliquorum omnium eloquentissima commentaria admirari atque amplexari, modo non negent hunc librum habere sua dona, suum genium, neque linguae scabriciem nos tanto odio prosequemur." The original that Heseler cites is from Dryander, *Anatomia mundini*.

69. See French, "Note."

70. Heseler, *Andreas Vesalius' First Public Anatomy*, 52–55: "Quod autem Mundinus dicit se non alto stilo uti in his velle, certe non intelligo quid per stilum intelligere velit. Certe Mundinus non observavit hic ordinem Galeni. Nam stilus et via anathome est sequi modum et ordinem in ea secundum Aristotelem et Galenum. (Certe ille bonus Curtius huiusmodi hic non intellexit, quid scilicet proprie vocetur stilus. Significat enim modum loquendi, et sententia et mens Mundini fuit, se in his non velle uti exornatis et comptis verbis sed simpliciter velle anathomen tradere familiari sermone. Res enim contenta est doceri, et magis dilucidius quam elegantiis tradi debere. Herbis seu rebus, non verbis curamus. Recte igitur dixit Mundinus.)"

71. Ibid.

72. Mondino, *Anothomia*, 96: "Hinc est quod inter cetera vobis cognitionem partium corporis humani, quae ex anothomia insurgit, proposui tradere, non hic observans stilum altum, sed magis secundum manualem operationem vobis tradam notitiam." The English translation is by Charles Singer, who, in addition to translating the 1493 version

of the *Fasciculo di Medicina*, also provided a translation of Mondino's Latin text in that work. See Ketham, *Fasciculo di Medicina* (trans. Singer), vol. 1, 59. The 1493 version of the *Fasciculo* states: "e perche la cognitione delle parti del subiecto nela medicina e el corpo humano el qual si chiama li luoghi dele dispositioni e una delle parti della scientia dela medicina. e de qua nasce che fra tucte laltre cose dovemo haver cognitione del corpo humano e delle parti de esso: la qual cognitione insurge e procede dalla anathomia. la quale ho preposto de dimostrare: non observando stile alto: ma secondo la manuale operatione vene daro notitia."

73. This was an argument that gained traction in the language debates of the period and in ideas about the utility and success of the vernacular (as opposed to Latin). By the early seventeenth century, the anatomist Giulio Casseri would recommend the nude (or naked) style of speech, and although Casseri meant it to describe Latin, many other writers from that period—Ambroise Pare, Leonardo Fioravanti, Thomas Sprat—connected such nudity to the rising significance of the vernacular. See *Vulgariser la médicine*, esp. 8–9; and Holz, "Le style nu."

74. The issue of style remained in the commentaries and translations of Mondino's work. See *Questyonary of Cyrurgyens*, which is the basis for Thomas Vicary's ideas on verbal clarity; and Vicary, *The Englishe Mans Treasure*. See also Lanfranco's work on surgery, *Chirurgia parva Lanfranci*. Moreover, in his bloodletting manual, Pietro Paolo Magni explained that the surgeon should speak clearly and logically, repeatedly cautioning against *sporche parole*, literally "dirty words." See Magni, *Il modo di sanquinare*, both in print and in a manuscript (ms. Cod. Cic. 277c), based on the printed edition, held in the Library of the Correr Museum, Venice.

75. Colombo arrived in Padua in 1538 and held the chair of surgery from 1541 to 1544, when he left to take a position in Pisa. He resuscitated the ancient anatomical projects of the Alexandrian anatomists Herophilis and Erasistratus, exploring (as they did) vivisection and what it could elucidate about anatomical matters. See Cunningham, *Anatomical Renaissance*, 143–66, esp. 147.

76. Colombo, *De re anatomica*, bk. 2, ch. 5, 184–85: "De asperae arteriae cartilaginibus: Neque mirum videri debet, tot celebres viros hallucinatos esse in substantia laryngis: facilis enim est error, si brutorum, ut bovis, et huiusmodi aliorum laryngem intueare, qualem Vesalius publice semper secare, atque ostentare conseuvit. Verum si humanam inspexeris, praesertim in consistente aetate, procul dubio ossicula omnia, ex quibus conficitur, agnosces epiglottide excepta, quae fistulam tegit."

77. Ibid., bk. 5, ch. 20, 255: "De musculis thoracem moventibus: Exteriores enim sursum trahunt, interiores deorsum, atque ita trahunt costas simul, validemque constringunt: nam vis constringendi in thorace valida esse debebat, eo quod foras expiramus, et vociferamur: et hunc motum ego saepius in vivi canis sectionibus observavi, sectionibus, inquam, quae domi fiunt, non quae palam: ibi enim omnia, ut in privatis sectionibus, exquisite considerari nequeunt, ob auditorum spectatorumq; frequentiam."

78. Ibid.: "Saepe enim tercentis, et amplius auditoribus in Academia patavina et Pisana, et Romae deniq; ubi iam decem annos profiteor, me circundatum vidi. Causa quam obrem Vesalius in hoc hallucinatus fuerit, haec fuit: quod existimaverit in hoc motu primam costam immotam permanere."

79. According to Moritz Roth, Vesalius left Padua in the first months of 1544. See Roth, *Andreas Vesalius Bruxellensis*, 187. In 1545–46, Columbo occupied the chair of surgery. See *Rotoli*, filza 651, f. 173v, Archivo Antico dell'Università di Padova (hereafter cited as AAUP). It was decreed on 12 December 1547 that Giovanni Paolo Guidaccio, one of Vesalius's students, would assume the role of demonstrating publicly; the decree noted that Guidaccio had been doing so "in many places publicly and to the satisfaction" of his audiences. On 14 December 1550, Alessandro Sarego was assigned to give the lectures on surgery, and on 14 December 1550, he gave the anatomy and then the reading, or lecture, that lasted until 15 February 1551. See Pietro Tosoni, *Della anatomia degli antichi*, 87; and *Raccolta Minato*, n. 646, AAUP. As Giuseppe Favaro explained, the roster of chairs (or *rotolus*) lists the chair held by Falloppio as *ad Cyrugiam et Lecturam Simplicium* and the *Bollettari degli stipendi* lists his position as *semplici, chirurgia, et obligo di (tagliar la) notomia*. In 1551–1552 and again in 1559–1560, the Senate's deliberations on Falloppio's salary list his responsibilities for "reading" in surgery, for simples, and for cutting, reading and showing the anatomy "as in years past." See G. Favaro, *Gabriele Falloppia modenese*, 85–106.

80. *Atti dell'universita artisti*, 1434–36, 1531–1557, filza 675, f. 165, AAUP, transcribed by G. Favaro, *Gabrielle Falloppia modenese*, 223: "Magnifici et Clarissimi DD patres et huius patavini Gymnasii Instauratores dilligentissimi: Havendo nui inteso qual sia lo animo et intentione di Vostre Clarissime Magnificenze circa lo administrar de la anothomia atiò pienamente si adimpiscano li statuti di la nostra università habiamo convocata la bancha et congregatione nostra, dove per fiat è stato determinato che lo Eccellente messer Vector Trinchavella deputato alla lectura di la pratica ordinaria di medecina in primo loco, habbi a leger lui la anothomia questo anno, et che lo Ecc messer Gabriel Falloppio habbi a tagliar et mostrar solamente, si come si contiene nelli predetti statuti. Perhò essendo necessario che questa nostra ellectione sii confirmata da Vostre Clarissime Magnificenze ne è parso scriver la presente a V. CL. Signorie pregandole humilmente che voglino dignarsi di confirmare la detta nostra ellectione, essendo questo universal beneficio di tutto il studio, facendo intender per lettere sue, a gli sopradetti doctori, che cossì vogliano exequir quanto per nui è stato determinato, et alla buona gratia di Vostre Clarissime Magnificenze senza fine si racomandiamo. Ex offitio Artistarum Paduae Die 6 mense Xmbris MDLIIII. E. M. V. Deditissimi Artistarum Vice rector et consiliarii."

81. *Atti dell'universita artisti*, 1434–36, 1531–1557, filza 675, f. 171, AAUP, transcribed by G. Favaro, *Gabrielle Falloppia modenese*, 227: "Clarissimi Signori. Sendo stato per vostre magnificentie clarissime ordinato che questo anno si dovessero partire li carichi in fare l'anatomia: si era eletto l'Eccellente Apellato a leggere il testo del Mondino: e l'Eccellente Trincavella ad esporlo, e mostrare: e l'Eccellente Faloppio a tagliarla. Hora quando si è venuto al fatto, in tutto si è osservato l'ordine, ecetto che l'Apellato non ha fatto quello che gl'era stato imposto, di leggere il testo del Mondino: e questo perche cosi è parso meglio al Trincavella: quale cominiciò a leggere nelle schole, e gli lesse [per] due [gi] lettioni la prima assai quiete, la secondo più presto interrotta ch'altrimente. Venne di poi a leggere nel Teatro e con gran sua difficultà fu la prima volta udito, e venendo al mostrare li scolari non [ne] rimasero [molto mal] troppo sodisfatti; la seconda fu del tutto interrotto, nè mai puotè dir parola per il rumore de scolari che gridavano vogliamo il Falop-

pio. E per quanto pregare gli facesse il Faloppio che stessero quieti, e che non dubitassero che gli saria stato dal Eccellente Trincavella sodisfatto, sempre gridavano vogliamo il F. E cosi la lettione andò in nulla."

82. Ibid.

83. Students become a consistent part of the modifications and reforms of anatomical lessons in the second half of the sixteenth century. *Atti dell'universita artisti*, 1434–36, 1531–1557, filza 675, f. 171, AAUP, transcribed by G. Favaro, *Gabrielle Falloppia modenese*, 221–222: "Al capitanio di padova sul mutamento d'orario della lettura di chirurgia, 8 March 1552: Intendemo che li scholari che vogliono udir l'eccellente messer (conte da monte) Gabriele Falloppio non si contentano andare ad udirlo alla 19 overo alla 22 hora che alla lettione di chiruggia è assignata dal statuto, imperochè alla 19 è troppo presto doppo il pranso ed alla 22 legge l'eccellente Geneva, ma che si contentariano che legesse mentre sona la campana ciò è alle 20 in 21 hore. Per la qual cosa havemo fatto le presenti alla magnificetia vostra [che] et la pregamo che havendo rispetto al desiderio di scholari, et alle lettioni di quegl'altri eccellenti dottori, la voglia trovare quel hora che possa tornar meglio a cadauno, la qual remettemo alla prudentia della magnificenza vostra ponendo poi ordine che all'ora che haverà deputata la magnificenza vostra detto eccellente messer Gabriele Falloppio come habbia a leger la detta lettione de chirugia." Favaro also gives additional examples on pp. 221–222.

84. *Atti dell'universita artisti*, 1434–36, 1531–1557, filza 675, f. 171, AAUP, transcribed by G. Favaro, *Gabrielle Falloppia modenese*, 221–222: "L'altro giorno medesimamente la mattina non puotè leggere se non con grandissimi strepiti, nè fu udita parola alcuna: poi la sera non volsero ch ea niun patto leggiesse, chiamandosi tutti mal sodisfatti. E così l'anatomia è stata risolta. Però vedendo noi il danno grande che al Studio aviene per non farsi detta anatomia in questo Studio, che è il più famoso di tutto il mondo: tanto piu che molti scolari per essa sono qui rimasti in fino ad hora, e molti anchora da altri studii per questa solo sono venuti: pregamo Vostre magnificentie clarissime che ci vogliano concedere che faciamo l'anatomia secondo il modo de gl'anni passati, massimamente che veddiamo l'Eccellente Faloppio essere da tutti chiamato; il quale solo sempre ha meglio sodisfatto, di quello habbiamo visto questa moltitudine haver fatto, e che possiamo sperar che mai debba fare; la quale come habbiamo visto, più presto è stata causa di confusione, che di ordine alcuno. Si che per le cause predette, e perchè anchora il tempo il permette, et oltre il clarissimo Podestà ha un corpo da giusticiare quale [ha] ci ha promesso de novo supplichiamo Vostre magnificentie clarissime, che in questo voglino sodisfare al commune desiderio et utile di questo Studio."

85. Ibid.

86. French, "*De Juvamentis Membrorum.*"

87. Galen, *On the Usefulness of the Parts*, bk. 17, 724–28. See also Temkin, *Galenism*.

88. Park, "Organic Soul."

89. Heseler, *Andreas Vesalius' First Public Anatomy*, 55–60.

90. The relationship between anatomy and clinical medicine in the early modern period is not fully understood. See Carlino, *Books of the Body*, 139–43.

91. See n. 87.

92. Carlino, *Books of the Body*, 143.

93. Ibid.

94. It was assumed that more information about anatomical structure would yield more information about function, leading anatomists to focus on structural composition. See Bynum, "Anatomical Method."

95. See Historia: *Empiricism and Erudition*, esp. Pomata, "*Praxis Historialis.*"

96. Pomata, "*Praxis Historialis,*" 111. In this, Falloppio followed not only Galen but also the approaches of Berengario da Carpi, Benedetti, and Massa, as well as Vesalius. Jacopo Berengario da Carpi referred to his method as *anatomia sensibilis*; Alessandro Benedetti used Greek examples that were drawn from experience (or indirect observation); and Niccolò Massa called his method *anatomia sensata*. This was the tradition that Vesalius called forth when, in the dedication of *De humani corporis fabrica*, he emphasized the importance of seeing with your own eyes. See also Kusukawa, "Uses of Pictures."

97. Falloppia, *Observationes anatomicae*, 63. The first volume of this work is a facsimile reproduction of the first edition of Falloppio's *Observationes anatomicae* (Venice: Marcum Antonium Ulman, 1561). Unless otherwise noted, all quotations are taken from the first volume. All translations are my own.

98. Ibid., 86.

99. Ibid., 8: "De usu autem appendicum non loquar, cum ad anatomicam speculationem, quam in opere maiori perficiam, pertineat haec tractario. Sed unum hoc tibi promitto, me ibi esse verum appendicum usum proditurum, quod a nullo quantumuis docto anatomico adhuc factum est."

100. Ibid., 86r–87v: "Audio quosdam, qui se mihi amicus profitentur, publice hanc opinionem aliquando irridere; atque rationes etiam aliquot afferre, quibus probent auditoribus suis hos non esse musculos distinctos, sed potius rectorum partes. Nam asserunt, Particulae iste musculorum definitionem minime habent; ergo musculi distincti non sunt. Secondo, Carnes istae aliquando non reperiuntur; ergo quintum par musculorum à natura institutum minimè faciunt. Quoniam si hoc esset, semper reperirentur. Tertio, Nullum habent usum proprium; ergo musculi non sunt, confirmatur autem ab his nullum habere usum carnes istas. Primo, quia non faciunt ad priapum ipsum erigendum; quoniam nullam cum ipso iunctionem habent. Secundo, quia nec vesica, nec aliquod aliud viscus in inferiori abdomine contentum ab his conprimi potest, cum satis valide ab omnibus octo reliquis musculis comprimantur. Tertio, non credendum est, quod rectis opem ferant, cum ita magni validique ii sint ut nullo adiumento praecipue istarum minimarum carnium egeant. Quam obrem ex his omnibus sequitur (ut ipsi afferunt) carnes has quintum musculorum par minime conficere; sed esse potius rectorum particulas. Doleo ego, quod isti virium suarum non sint aequi aestinatores; quoniam non ita libere in alios anatomicos loquerentur, sed per me liceat quicquid in me velint; amicitiae enim nostrae illud dono; atque contentionis istius occasionem laudo. Quoniam . . . ut ait Poeta, semperque ex huiusmodi litibus aliquid in commodum illorum, qui addiscere volunt cedit."

101. Ibid., 87r: "Nam cum primo dicunt, quod definitionem musculi minime habent, ingenue fateor me ignorare quid sibi velint per definitionem musculi. Quoniam si definitionem intelligunt, quae tragelaphum nobis exprimat, haec procul dubio istis minime

competit. Verum si illam agnoscunt, quam ego, et ante me Galenus in principio primi de motu muscu. agnovit. Musculum scilicet esse immediatum motus voluntarii organum. Dico hanc definitionem sumopere ipsis quadrare, esseque voluntariis motus vera organa, atque suminis partes voluntaria compressione oblique deorsum trudere et consecutioni certa musculos esse, atque a me ideo appellari."

102. Ibid., 88r–89r.

103. Ibid., 65v–66v: "Anno ab ortu Christi millesimo quingentesimo quinquagesimo tertio, quarto kalendas Ianuarias, dum hic Patavii in publica anatome totus essem, Matthias Guttrich medicus Germanus (vir non solum philosophorum ac medicorum disciplinam, et latinae, graecaeque ac hebraicae linguae cognitionem doctus verum etiam in penetrandis medicamentorum medullis per ignis vim, atque sublimationem vocatam ita excellens, ut omnes, quos ego adhuc noverim penitus superet) ad mirabili quadam ratione Venetiis medicinam faciens, ad me misit caput phocem in foro piscario emptum, detulit autem Rubertus Phinch tuem auditor meus, ac veluti filius amantissimus, qui postquam in excellentem philoshopum [sic] et medicum evasit, artem medicam nunc exercet summa cum laude et gloria . . . Hoc acceptum caput secare cepi, atque secando observavi animal hoc utramque palpebram movere, et oculum unde quaque detegere. Instrumenta erant quatuor musculi rubri admodum in oculi orbita una cum reliquis ocularibus musculis latentes et cum ortu illorum, qui motuum rectorum oculi opifices sunt connati; unus in parte superiori; alter in inferiori; tertius in interno angulo, quartus in exteriori situs erat . . . Quam structuram publice in theatro ostendi, atq; inde admonitus, statim in bovino oculo simile instrumentum, quo superior tantum palpebra tollitur, disquirere cepi, illudque unicum inveni. Ab hoc exemplo doctior factus, etiam in humano oculo reperi musculum parvum et tenuem admodum, cuius principium ab eodem penitus loco oritur, unde etiam manat principium musculi oculum recta ad superiora attollentis." The episode is also mentioned by G. Favaro, *Gabrielle Falloppia modenese*, 89.

104. Falloppia, *Observationes anatomicae*, 6–7: "Ad lectorem: Sed quoniam quam plurimi reperiuntur, qui aut scribendo, aut publice, privatim ue profitendo, multa invenissem quae ab aliis minime visa sint in hac arte sibi arrogant, seseque; alienis plumis ornantes, uti Aesopeus graculus, ea quae ab aliis, aut a me ipso, vel potius a meis discipulis audientes male didicerunt, pro suis edunt, et venditant; et aliquando ita sibi arrogant, ut eadem distorta et corrupta exponant."

105. Ibid., 61r–62v: "In hoc capite, nisi historia me retraheret, vellem quod disputationem quandam legeres, qua constituo quid sentiendum sit de unitate, aut multitudine musculorum; cum in hoc anatomici valde inter se discordes sint; addoque canonem, quo unusquisque musculus cognosci potest an alterius pars sit, et an potius per se distinctus consistat. Sed quoniam in ea tractanda prolixior sum, quam hic locus requirit, ideo disiunctam ab hac historia, cum novum dabitur otium ad te mittam. Musculosque nunc aggredior, inter quos illis, qui occiput tenent, primum lucoum ascribam. Secundum musculis aurium. Tertium frontis, ac postea reliquis suum servato eo ordine, quem Vesalius sequutus est."

106. Ibid., 98r–99v: "Dorsi totius musculi ita varii et complicati sunt, ut non sit mirum si anatomici scriptores inter se concordes non erunt. Nam ut quid sentiam ingenue

profitear indigesta moles, atque confusum chaos musculorum mihi videtur, in quo prae-
ceptorem desidero, qui distincte ante oculos hos mihi dissecet, ipsiusque partes ad cer-
tum numerum ac ordinem di ducat. Minime enim in his musculis explicandis, quod in
paucissimis aliis accidit, mihi satisfacio, eoque minus cum videam, quod si ego voluerim
meliorem instituere divisionem ob rei difficilem naturam, alius fortasse anatomicus eam
assequi non poterit ob infinitam fibrarum et originum et insertionum, quae in hoc chao
continentur, multitudinem. Est enim veluti plurimarum viarum labyrinthus, in quo tamen coram te quid observarim libere dicam."

107. Ibid.

108. See n. 6

109. Fabrici, *De motu locali animalium*, 82: "Miraberis forsitan lector, quod musculos
non describam, ut Vesalius in toto suo opere, et Galen. in libro de adm.anat.fecit, qui
ordinem, seu commodam dissectionem respicientes eos descripsere, quoniam ii tan-
tummodo eorum dissectionem pro ut unus alter i succedit, et contiguus est associaturq;
nobis saltem ob oculos ponere, et monstrare voluerunt. At nos, qui scopum habemus
docere per ea, quae insunt musculis, eorum actiones et usus, merito alio ordine in ce-
dendum duximus, qui procul dubio nos ducit ad notitiam casuum musculorum et ar-
ticulorum. Nam si quis simplicem dissectionem inquirat, et primum secundum tertium
et sequentes hoc modo numeret, potius confusionem, quam notitiam utilitatem musculorum consequetur. At quando nos eorum, quae insunt musculis, causas inquirimus,
tunc usum inquirimus, et musculorum numerum exactius memriae mandamus."

110. To study the chick embryo, Fabrici began with what were basically seven structures, though the uterus signified three of them: the uterus, the egg, the vitellarium (a
group of yolks) or raceme (the ovary, a group of eggs), the pediolus (the stalk to which
yolks are attached), the peduncle (attaching the yolk to the stalk), the infundibilum (the
passage between the first and second uterus), and the podex (where the second uterus
terminates). See his "De formatione ovi et pulli" in *Embryological Treatises*, 141–48. In *De
formato foetu*, Fabrici limits the structures to nine. See "De formato foetu" in *Embryological Treatises*, 247–75.

111. Cunningham, "Fabricius and the 'Aristotle Project'"; and Cunningham, *Anatomical Renaissance*, 167–90.

112. Pomata, "*Praxis Historialis*," 117.

113. Scholars also used to describe Fabrici's transition as one from human to com-
parative anatomy. See the introduction to *Embryological Treatises*. As John Herman
Randall, Jr., has argued, the debate on the immortality of the soul engaged many of
the faculty at Padua; it depended on several Aristotelian texts and conflicted with
both Catholic and Protestant doctrines, which held that the soul was immortal. See
Randall, "Development of Scientific Method"; and Pagel, *William Harvey's Biological
Ideas*, 19–20. On the materiality of the soul, see Cunningham, *Anatomical Renaissance*,
170–71.

114. See Fabrici's "De formatione ovi et pulli" in *Embryological Treatises*, 86.

115. For example, the graphic schemes that delineated the commentary tradition
provided an organizational structure for the body of anatomical knowledge and a way to
remember parts of it. See Rinaldi, *Arte sinottica*.

116. On the role of smell in the emergence of medical dispassion, see Payne, *With Words and Knives*, 17–26.

## CHAPTER 2: FABRICI'S DOMINION

1. These interim anatomists included Prospero Borgarucci (b. 1540) and Nicolò Bucella. On Borgarucci, see Rinaldi, "Modèles de vulgarisation." For Bucella, whose family was associated with the Anabaptists, see Stella, "Intorno al medico padovano." For the biography of Fabrici, see *Embryological Treatises*, 1–32. For a thorough account of his teaching, see G. Favaro, "L'insegnamento anatomico." From 1565 to 1613, Fabrici was a professor of surgery and anatomy at Padua and, from 1600 on, he was a professor supraordinarius in anatomy. He prepared his research for publication between 1600 to 1619, when he died.

2. *Atti della nazione germanica*, 1560, vol. 1, 56: "Die XVIII Decembris D. Hieronymus Fabricius de Aquapendente, chirurgiae professor, anatomen auspicatus est, quam pro viribus non citra scholasticorum applausu administravit, eamque die 5 Ianuarii anni 1567 absolvit."

3. Ibid.

4. For example, Vesalius is praised throughout Baldassar Heseler's notes. See Heseler, *Andreas Vesalius' First Public Anatomy*. Similarly, Falloppio is praised in *Atti della nazione germanica*, 1560, vol. 1, 34–35. See n. 5. Roger French has referred to this support as the "jury effect" and suggested that the legal construction helped to confirm the truth of the anatomist's claims, as well as verifying the visible evidence presented during a demonstration. See French, *Dissection and Vivisection*, 200–206. On the genesis of facts in legal settings, see B. Shapiro, *Culture of Fact*, 8–33.

5. *Atti della nazione germanica*, 1560, vol. 1, 34–35: "Habuimus ergo hoc anno ex sententia Anatomicam satis diligentem et accuratam, exquisitiorem forte habituri nisi et tempus aliquantisper contrarium nobis fuisset, et Fallopius in medio quasi deficiens minime firma uteretur valetudine, quo nimirum factum est ut postea ad finem nimis fuerit properatum."

6. Ibid., 1568–1569, vol. 1, 63: "Anatome tum administrabatur duorum cadaverum, viri et mulieris, peducisque vivae, si non exquisitissima, saltem tolerabilis, praestita etiam reliquis honorificentissima sepultura." The passage is briefly discussed by G. Favaro, "L'insegnamento anatomico," 107–36, esp. 125.

7. *Atti della nazione germanica*, 1568–1569, vol. 1, 63.

8. Ibid., 1572, vol. 1, 84–85: "et hoc vel simili responso (ut in Universitate postmodum convocata accepi) substitutum a se dimiserunt, obligari asseverantes Dominum Hieronimum ab Aquapendente singulo quoque anno ad anatomen publice administrandam neque ob aliud adiectum fuisse nuper ad ipsius salarium. Subolfecit is autem voluntatem superiorum: quare commutato consilio, etsi tempus cadaverum sectioni accommodum prope abiisset, sine mora tamen loco privato compendiosam nec plane sterilem aggressus administrationem et suo muneri et utilitati nostrae ex parte fecit satis. Verum huc quoque pertinet, quod Excellentissimus Dominus Buccella chirurgus, Nationis Germanicae studiosissimus, ubit constabat brumales ferias absque publica anatome frustra

scholaribus abire, interpellatus primum a me ipso, deinde et ab aliis, omnem operam suam ab benevolentiam in negocio anatomico detulit, paratissimum se aiens cadaver aliquod in nostram gratiam secare, dum id ipsi a nobis suggeratur." See also ibid., 1574–1575, vol. 1, 96: "Privatam vero, quo utilitati nostrae consuleret, se hoc anno aggressurum, iam in publicis lectionibus non semel sponte promisisset, quam etiam, licet satis tarde, sub finem mensis Ianuarii (quod minus firma ad id usque tempus valetudine uteretur) admodum concisam et non adeo exactam edidit."

9. G. Favaro, "L'insegnamento anatomico," 107–36.

10. Atti della nazione germanica, 1570, vol. 1, 69–70: "Fracti autem turbulenti eorum conatus ac Brixianis totius negotii cura concredita, qui strenuam plane navarunt operam, ut res tum tranquille, tum iucunde etiam perageretur, nec idonea deessent corpora, unde factum est diligenti administratione clarissimi viri D. Hieronymi Fabritii ab Aquapendente professoris chirurgici fructuosam omnino hoc anno habuerimus anatomiam publicam; qua absoluta, ut corporibus dissectis et suus daretur honos, funus institutum plane honorificum, quod Rector ipse sua praesentia ac plaerique doctorum ornarunt." See also ibid., 1572–1573, vol. 1, 92: "In eodem conventu et de anatomica tractatum est, quae sicut a Domino Hieronymo Fabricio ab Aquapendente publico chirurgiae professore ob sequentem annum, quo iubileus dicetur, lectionibus non dedicatum, nobis promittebatur, ita etiam exhibita est, satis diligens et acurata, et acuratior fortasse adhuc fuisset, nisi humida coeli constitutio intervenisset."

11. Cunningham, "Fabricius and the 'Aristotle Project.'" This line of thought is elaborated in Cunningham, Anatomical Renaissance, 167–90.

12. Fabrici's publication project, the Theater of the Whole Animal Fabric, contained colored illustrations. These emphasized the structural detail of specimens, and I suggest in chapter 3 that they, rather than dissected corpses, could serve as the main object of study, as mnemonic devices and as directional/spatial maps. On these illustrations in relation to the three fundamental categories of Renaissance Galenic anatomy—historia, actio, and utilitas—see Siraisi, "Historia, actio, utilias."

13. Philosophy was conjoined to medicine as early as the thirteenth century, via the works of Taddeo Alderotti (ca. 1215–1295) in Bologna and Pietro d'Abano (1257–ca. 1315) in Padua. See Siraisi, Taddeo Alderotti; and Ottosson, Scholastic Medicine and Philosophy, 68–88.

14. Aristotle, Metaphysics, I.1; and Aristotle, De anima, III.3. See also Ottosson, Scholastic Medicine and Philosophy, 88–90.

15. While this chapter considers the pedagogical culture more fully, as well as the ways in which Fabrici assimilated Aristotle and Galen, it builds on the work of Andrew Cunningham, who has described and explained Fabrici's interest in Aristotle.

16. Atti della nazione germanica, 1575–1576, vol. 1, 101: "Instante bruma, publico scholae queritu decretum est, ut Hieronymus Fabricius chirurgiae professor, commodam in hanc rem tempestatem nactus, publice humani corporis anatomiam faceret. Eam ille anno insequenti ad octavum idus Ianuarii auspicatus, VII cal. Februarii ad finem perduxit, pluribus consectis cadaveribus tanto studio tantaque industria ut exactam hinc absolutamque humani corporis fabricae cognitionem, qui tantum non lapis [sic] spectavit, facilime hauserit." [There is a marginal note here that reads "What?"]

17. See n. 8.

18. *Atti della nazione germanica*, 1577–1578, vol. 1, 132–33: "altera erat morbus Eccellentissimi nostri Anatomici, qui maximam hyemis partem aeger oculis sic affectus erat . . . Eccellentissimus Aquapendens per valetudinem praesse sectioni nequeat . . . [after the Rector insisted] Eccellentissimus Anatomicus tandem promittere, se potius quam alteri provinciam hanc tradi patiatur, etiam cum detrimento valetudinis anatomen habiturum domi suae, ita ut in conclavi suo unum post alterum secaret membrum et spectantibus exhiberet. His tractationibus unus et alter dies consumptus . . . Cumque huic negotio se admiscuissent primarii aliquot ex nostris Consiliariis, tota spes anatomes evanuit." Giuseppe Favaro discusses this as an example of Fabrici's tendency to postpone demonstrations. See G. Favaro, "L'insegnamento anatomico," 114–15.

19. *Atti della nazione germanica*, 1577–1578, vol. 1, 132–33.

20. For a discussion of Galen's three categories, see chapter 1. On Fabrici's *historia*, see Pomata, "*Praxis Historialis*," esp. 117; and Maclean, "White Crows, Graying Hair."

21. Cunningham, "Fabricius and the 'Aristotle Project.' "

22. Fabrici, *De formato foetu*, pt. II, ch. 1. The translation is from *Embryological Treatises*, 276.

23. Fabrici, *De motu locali animalium*, 3: "Toto hoc tempore, quo non popularem, sed exactam anatomen administramus, agere in vestram gratiam, auditores, divino favente auxilio constitui de motu, quo totum animal loco movetur: seu de motu locali totius animalis, seu mavis dicas, de motu, quo totum animal locum, sue positionem mutat." The passage is also cited in Cunningham, "Fabricius and the 'Aristotle Project.' "

24. Cunningham, *Anatomical Renaissance*, 176–77.

25. Park, "Organic Soul."

26. Cunningham, *Anatomical Renaissance*, 176–77.

27. Fabrici, "De visione," in *Opera omnia*, dedication: "Quod si ex tribus hisce partibus prima dividatur; ut aucto iam earum numero fiant quatuor, Dissectio, Historia, Actio, Usus: nihil reprehendo. Video enim et veteres quosdam id fecisse, et fieri non incommode posse; cum re vera, exacte si distinguas, aliud sit Dissectio, aliud Historia. Dissectio primum rationem dissecandi organi tradit, quotque modis ea fieri et possit, et debeat, ostendit. Est autem ad hunc finem comparata, ut reliqua tria, videlicet historiam, actionem, et utilitatem patefaciat. Historia autem ea exponit, quae organo insunt, nimirum, temperamenta quae consequuntur, quaeque accidunt; quae omnia aperiuntur dissectione; qua eadem innotescit etiam Actio, ut et Galenus testatur, et nos in Anatomica methodo demonstravimus. Est autem Actio uniusc

iusque organi munus seu functio. Utilitas vero nihil aliud, quam id, cuius gratia tributa cum ipsa actio, tum ea quae, insunt."

28. Fabrici, *De formato foetu*, pt. I, ch. 1. The translation is from *Embryological Treatises*, 247.

29. Fabrici, *De venarum ostiolis*. This version includes a facsimile of the original text and a translation.

30. *De venarum ostiolis*, 70 [from the facsimile]: "Cogitanti mihi iam dudum, cui potissimum, tanquam benevolo et fautori, hunc meum de Venarum Ostiolis tractatum dicarem, nullus succurrit, cui magis eum convenire existimarum, quam Inclyte Nationi Germanicae; ut quae inter ceteras hoc meum de iisdem Ostiolis inventum prima mecum

observarit, mecum in sectione corporum iucunde contemplata sit, mecum admirata, vos is estis, qui preter ceteros Anatomen expetistis."

31. Ibid., 72 [from the facsimile]: "Contra vero quispiam priores in re hac insimulaverit, quod usum ostiolorum, qui apprime videtur necesarius indagare neglexerint, quodue ipsa in venarum ostensione non animadverterint. Nam nudis venis, iisq; integris ante oculos oblates ostiola se se quodammodo in conspectum exhibent."

32. Fabrici, "De formatione ovi et pulli," in *Embryological Treatises*, 86. Cited by Cunningham, "Fabricius and the 'Aristotle Project,'" 210–11.

33. See *Il teatro dei corpi*.

34. Capivacci, *De methodo anatomica*. This was published again as *Methodus anatomica, sive, ars consecandi* (Frankfurt: Aere et Operis Egenolphianis, 1594). The *Dictionary of Scientific Biography* (hereafter cited as *DSB*) indicates that it was formulated between the 1560s and the 1580s.

35. Capivacci, *De methodo anatomica*, 3: "non modo circa dissectionem versari, verum etiam circa actiones et usus . . . Artis anatomicae haec sit definitio. Ars Anatomica est ars naturae hominis sectione, actione, et usu comparata." According to the *DSB*, this treatise was formulated earlier, alongside other methodological statements, all of which were subsequently published in the 1590s.

36. Capivacci, *De methodo anatomica*, 4–5: "Declaro, loco forme et proinde generis dictum fuit, esse scientiam, ut distingueretur a peritia, quae pertinet ad incisiores, ad manuales . . . Loco autem subiecti, et proinde differentiae dictum fuit. (4) hoc est non modo sensu, sed etiam ratione comparatum; Quantum ad brevem expositionem, sciendum hanc traditam esse per formam, subiectum, et finem, omittitur autem causa efficiens, quoniam ex necessitate supponitur, ut si proponatur superva canea videatur, quoniam non est in causa, ut hoc definitum a quocunque alio magis distinguatur quam ex particulis propositis."

37. The *demonstratio propter quid*, which imparted information about causes, was seen to develop from the *demonstratio quia*, which was the description of the part.

38. Stephen Gaukroger, *Emergence*, 118–21. On the prehistory of these developments, see Lohr, "Metaphysics."

39. Fabrici's project placed anatomy in a more distant relation to medicine, but this was a familiar kind of move. Avicenna, whose *Canon* was a staple of medical education in the Renaissance, was interested in both theory and practice, but he made medicine subordinate to natural philosophy, which was more worthy than medicine because it interrogated the causes of change. On the fortunes of Avicenna's text, see Siraisi, *Avicenna*.

40. Although the university curriculum intended for a student to pass from logic to natural philosophy on his way to medical studies, Capivacci seemed to reverse the direction, arguing that anatomy was a philosophical endeavor and that anatomical knowledge as *scientia* could be an end in itself. Capivacci probably had the following dictum in mind: "Where the philosopher ends, there begins the physician." This dictum, drawn from Aristotle's *De sensu et sensato* and parts of *Parva naturalia*, was quoted by numerous Paduan philosophers and professors of medicine: Jacopo Zabarella, Cesare Cremonini, Giambattista Da Monte, Oddo degli Oddi, Fabrici, Giulio Casseri, and others. See Schmitt, "Aristotle," 12. Also see n. 13.

41. See Grafton and Jardine, *From Humanism to the Humanities*, 124–25. They see the emphasis on procedure as marking a shift from the intellectual innovation of the first humanists to the bureaucratic or administrative production of a classical education, that is, the full institutionalization of humanism that took place via a set of classroom aids—textbooks, manuals, teaching drills—and reduced humanism to a system. While their account suggests that charisma—the idea that humanism began as the practice of an exemplary individual—diminished as humanism became an institutional affair, William Clark has more recently shown that the form of institutionalization predicted the loss, enhancement, and transformation of academic charisma. He argues that the modern research university reached a "dynamic equilibrium" between its sphere of rationalization and its preservation and cultivation of professional charisma. See Clark, *Academic Charisma*, 17.

42. On medical humanism, see *Medical Renaissance*; Bylebyl, "Medicine, Philosophy, and Humanism"; and Nutton, "Rise of Medical Humanism." For anatomy and humanism, see Cunningham, *Anatomical Renaissance*; and Carlino, *Books of the Body*. On Padua and Venice, specifically, see Bylebyl, "School of Padua," esp. 349.

43. Nicholas Jardine, "Keeping Order," esp. 197–98.

44. Bylebyl suggests that the environment of humanist inquiry also resulted in the construction of the botanical garden and the anatomical theater. On the botanical garden, see the following and their accompanying bibliographies: Dal Piaz, "L'orto botanico"; and Dal Piaz and Rippa Bonati, "Design and Form." For botany and its relation to the business of apothecaries, see Palmer, "Pharmacy"; and de Vivo, "Pharmacies."

45. *Epistolario di Gabriele Falloppia*, 63: "Questi nostri Signori sono sul mutare tutto lo studio col porre de prattici alla Theorica, et de Theorici alla prattica, et credo che senza fallo vacarà il luogo di prattica a concorrenza del Trincavilla andando il Francanzano a Theorica, al qual luogo io dovrei pensare, et forsi che facilmente l'havrei, non di meno non vi penso havendo l'animo fisso così per la gran speranza datami da V S la qual prego che di gratia mi dia qualche avviso o dentro o fuori acciò che m'acquieti perchè non è la maggiore molestia al mondo che stare sospeso." The letter is dated 24 October 1561.

46. The bibliography on Paduan Aristotelianism is extensive. See the work of Charles Schmitt, especially his "Aristotle."

47. In 1571, Fabrici had postponed the annual demonstration because the temporary theater had not been built and a sufficient number of cadavers could not be found. See *Atti della nazione germanica*, 1571, vol. 1, 76. See also ibid., 1572, vol. 1, 85: "Quare commutato consilio, etsi tempus cadaverum sectioni accommodum prope abiisset, sine mora tamen loco privato compendiosam nec plane sterilem aggressus administrationem et suo muneri et utilitati nostrae ex parte fecit satis."

48. *Atti della nazione germanica*, 1572, vol. 1, 85: "Verum huc quoque pertinet, quod Excellentissimus Dominus Buccella chirurgus, Nationis Germanicae studiosissimus, ubi constabat brumales ferias absque publica anatome frustra scholaribus abire, interpellatus primum a me ipso, deinde et ab aliis, omnem operam suam ac benevolentiam in negocio anatomico detulit, paratissimum se aiens cadaver aliquod in nostram gratiam secare, dum id ipsi a nobis suggeratur. Mox vero et Itali torpidi illius ocii pertaesi, ac praesertim Massarii duo promptitudine Buccellae perspecta (ut lucro etiam ex mortuis

comparando inhiant), in aedibus ipsius sub natalitia festa anatomen apparent, quae sane non exiguum splendorem atque gratiam prae se ferebat, nisi partim incuria et perfidia dictorum ministrorum partim difficultate nanciscendorum cadaverum laboratum fuisse laborious." See also ibid., vol. 1, 89: "Extraordinariam vero anatomiae admnistrationem clarissimus D. Buccella Patavinus, anatomicae doctrinae peritissimus, circa publicae sectionis finem aggressus est, autore nobili quodam Bressano, quam tamen ob cadaverum inopiam et temporis incommoditatem imperfectam reliquit."

49. Premuda and Bertolaso, "Prima sede dell'insegnamento."

50. *Atti della nazione germanica*, 1578, vol. 1, 143–44: "Circa finem vero Octobris, cum coeli constitutio frigidior aliquanto esset, decreverunt mulierum quae in nosocomio illo morirentur, cadavera aperire et auditoribus locos affectos et morborum fomites demonstrare."

51. Ibid.: "id consilium feliciter satis ceptum in duobus corporibus subito fuit eversum. Cum enim die sequenti uteros harum mulierum aperire constituissent, et in altera, quae marasmo consumta erat et fistulam sub pectore habebat insignem, incidere et commonstrare quonam penetrasset fistula, Aemylius Campolongus ipsorum aemulus suis eodem die uteri sectionem pollicitus, uteros horum cadaverum abstulerat; unde factum ut re hac, et quaerelis anicularum idem, si morerentur, timentium, ad praefectos nosocomii delatis, interdictum sit et Oddo et Campolongo, sub poena amissionis salarii, ne quod cadaver in posterum aperirent."

52. On the relationship between dissection and disrupted burial, see Park, "Life of the Corpse."

53. *Atti della nazione germanica*, 1578–1579, vol. 1, 145–46: "Ita, controversia illa sopita, agi ceptum est de anatome, quae tamen in 16 usque Ianarii diem dilata est, tum propter coeli constitutionem minus commodam, tum propter rescriptum Clarissimorum Reformatorum, quo vetabant ne ante eum diem initium anatomies aut finis lectionum fieret, propter intervallum inter natalitia festa et quadragesimam insigne."

54. The medieval origins of this tradition can be found in the dissections of nuns, done in search of the anatomical signs of sanctity. See Park, *Secrets of Women*, 39–76.

55. Girolamo Mercuriale published an eclectic assortment of works. See Siraisi, "History, Antiquarianism, and Medicine." Girolamo Capivacci published on the method of diagnosis and the method of anatomy; Marco degli Oddi published on women's diseases; and Emilio Campolongo published on method, extending Capivacci's work.

56. *Atti della nazione germanica*, 1576–1577, vol. 1, 108–9: "[4 January] ita tamen ut sine concurrentia (quemadmodum loquuntur) ambo Doctores Capovaccius nimirum et Mercurialis legerent. Hic quidem mane tractatum de pestilentia postea publici iuris factum: ille vero a prandio hora consueta morbos supremi ventris explicandos in manus sumeret: quae commoditas utrumque audiendi num ante hoc tempus unquam acciderit, valde dubito. Anatomica administratio ob caussas per se notas hac hyeme prorsus intermissa est. Post liberalia satis frigide celebrata, cum iam Hieronymus Mercurialis suas de peste commentationes absolvisset, Reformatores studiorum inclytii Gymnasii huius mandarunt, ut in posterum unus tantum, isque primarius atque senior tum theoricae tum practicae professor legeret, altero tamdiu quiescente, quo ad diversum super hoc negotio placeret. Quamobrem Clarissimus Paternus hora matutina solus aegrotantium

aliquot historias ex Hippocratis epidemicis explicavit; Capovaccius post prandium methodum suam curandorum morborum universalem (aliquoties ante etiam propositam) et retexuit et pertexuit; Dominus Mercurialis vero in quorundam gratiam pauca quaedam extraordinem docuit, quod et Clarissimus Anatomicus fecit."

57. Ibid. Capivacci, who took second chair in practical medicine in 1564, was described by students as "summus amicus et semper de eis optime meritus." The first chair was filled by Antonius Fracantianus Vicentinus. For the praise that the transalpine students lavished on both Capivacci and Mercuriale, see, for example, *Atti della nazione germanica*, 1576, vol. 1, 104: "maximo concursu et applausu omnium excepti sunt: verum fortuna ob vim veneni grassantis praesentaneam expectationi, quam de iis vulgus conceperat, non penitus respondente, parum abfuit, quin ignominiose ad extremum reiicerentur."

58. *Atti della nazione germanica*, 1578, vol. 1, 144: "De Excellentissimo etiam Anatomico compellando tractatum, quod tantum minus necessarium videbatur, cum ipse non tamen in publicis praelectionibus sed etiam in privatis colloquiis anatomen polliceretur et animos multa spe impleret et erigeret."

59. Ibid.

60. On these coordinated actions, see Cunningham, "Fabricius and the 'Aristotle Project'"; and Park, "Organic Soul."

61. *Atti della nazione germanica*, 1579–1580, vol. 1, 165 (transcribed after the interpolation of 1579, 146): "Quod ubi audivimus, communi consensu, Excellentissimos viros Hieronymum Capivaccium, Michaelem Cavassetum consulimus et repetitis verbis Aquapendentis, summis precibus rogamus, licet non publicam, privatam tamen anatomiam fieri. Sed viri isti nostris commodis et studiis addictissimi humaniter admodum respondent, se quidem nostris hac in re operam suam polliceri, verum dissentiente Hieronymo Aquapendente nihil quicquam moliri posse; vereri enim ne Aquapendens totum negotium si velit, impediri possit: quare sibi nihil videri consultius quam huius rei anatomicum prius admonere. Hoc igitur responso contentus, ex nostris Christophorum Homelium, Mathiam Kübergerum, Balthasarum ab Herden, Franciscum Hipolytum mihi adiungo ac vicissim anatomicum convenio, quem domi inventum placide rogo num aliqua spes futurae anatomiae reliqua sit. Tum statim nostrum quasi sermonem interrumpens, num preterito anno anatomiam vidissemus percunctatur. Nos vicissim multos propter hanc solam causam nuper advenisse dicimus, quibus gratissimum esset aliquam sectionem inspicere. Nobis itaque sic urgentibus nihil aliud ait, quam se vehementer mirari nostros lectioni suae privatae non interesse: nos contra horae incommoditatem in causa esse diximus. Ad postremum obstinatum ipsius animum perspitientes, ab eo licentiam alteri anatomiam palam tractandi petimus; verum illse alium praeter se hoc ipsum factitare posse negat. Fit igitur anatomia, sed ut coata, ita admodum exigua."

62. Ibid., 1579, vol. 1, 146: "Ad 16 igitur Ianuarii prooemium [*sic*] habiturus Excellentissimus Anatomicus, cui aderant etiam urbis Rectores, tanto strepitu et sibilis exceptus fuit, ut ne verbo quidem dicto cogeretur auditorio excedere, quae res Italos etiam ipsos dubios de anatome faciebat. Verum tranquilla omnia in posterum fuerunt et, 17 Ianuarii inchoata, anatomes administratio finita est 6 Februarii, consectis singulari industria pluribus tum virorum, tum mulierum cadaveribus, viva etiam anatome pecudis habita."

63. Ibid.: "et aderant frequentes aliqui ex doctoribus, qui antemeridiano tempore dubia proposita post meridiem solvebant."

64. Ibid: "Reliquiae cadaverum ad ecclesiam Cathedralem delatae et sepultae, comitantibus non tantum aliquot professoribus, sed scholasticis plurimis, inter quos nostri magno aderant numero."

65. Ibid., 1580–1581, vol. 1, 170–71: "Promiserat hic egregius vir a principio suarum lectionum auditoribus suis tractatum de tumoribus praeter naturam, de fracturis et (ut sui moris est loqui) forte etima de luxationibus. Sed, quod maioris erat momenti, pollicebatur nobis anatomen luculentissimam et absolutissimam. Tractationem de tumoribus praeter naturam quomodo absolverit scitis, doctrinam profecto de fracturis nunquam attigit, et tractationem de luxationibus omnino quoque omisit . . . Namque animum ipsius obstinatum neque praeceptorum intercessiones neque nostrorum preces frangere potuerunt, patientia sola nobis reliqua fuit quae et obscurissimam et imperfectissimam anatomiam loco luculentissimae et absolutissimae sustinere potuit." Also cited in G. Favaro, "L'insegnamento anatomico," 118.

66. Atti della nazione germanica, 1581–1582, vol. 1, 177: "Circa principium Ianuarii Anno 1582 in Universitatis conventu deliberabatur de epistola quadam nomine potissimum Nationis nostrae ad Venetos perscribenda in qua petebat Magnificus Dominus Vicerector a Venetis ut Dominum Aquapendentem singulo quoque anno publicam anatomiam exhibere cogant, aut saltem Universitati licentiam praebeant, altero anno, in quo vacat anatomia publica, aliquem sumptibus Universitatis eligere qui expectationi studiosorum satisfaceret, sed cum iam anatomia hoc anno neglectum fuit, in futuro sine dubio executio fiet. Hortamur itaque eos qui nobis successuri sunt, si mentio aliqua huius rei in Universitate fiet, ut diligenter id negotium urgeant." Although Fabrici promised to give public demonstrations annually, he did not keep that promise. The transalpine nation recorded his equivocations about giving the public demonstration as well as the regularity with which private lessons in anatomy were held. See also Atti della nazione germanica, 1581–1582, vol. 1, 177; ibid., 1583, vol. 1, 186; and G. Favaro, "L'insegnamento anatomico," 117.

67. Atti della nazione germanica, 1581–1582, vol. 1, 177.

68. Ibid.

69. Ibid.: "19 Ianuarii Anno 82 habita fuit anatomia privata in collegio Vicentinorum prope templum S. Catharinae, quae cum aliquibus Nationis nostrae non approbatur, semel atque iterum me convenerunt ac de anatomia privata inter nos facienda mecum deliberarunt, Nationemque convocari voluerunt, id quod sequenti die pro ac debui libenter feci. Convenimus omnes in aedibus meis, consilium et propositum omnibus placebat, omnes sumptus necessarios contribuere volebant. Excellentissimus vir Michael Aloisius chirurgus omnem operam et diligentiam nobis pollicebatur; in qua re satis ego quantum potui laboravi, ad quinque parochos ministrum publicum misi, qui omnes promittebant se prima quaequam cadavera oblata communicaturos, at quia propter penuriam demortuorum cadavera nulla acquirere potuimus, deliberatio nostra irrita fuit."

70. The series of decrees is recorded in Atti della nazione germanica, 1583, vol. 1, 192.

71. Raccolta Minato, n. 56, AAUP: "[5 February 1583] Et perchè ognuno sa di quantoa utilità, et honorevolezza sia al detto studio il far l'Anatomia ogn'anno, cosa che saepevolse

[*sic*] si è fatta per lo passato. L'andera parte, che'l D. Hieronimo Fabritii sia condotto a legger nello studio nostro di Padoa l'Anatomia ordinamente di anno in anno, et la Chirurgia insieme in questo modo però, che tutti li mesi dell'inverno sia tenuto leggere, tagliare, et mostrare l'Anatomia come lettura ordinaria, passati veramente i mesi d'inverno non possendosi per li tempi caldi maneggiare si corpi morti, che si puovefanno sia tenuto legger per ordinaria la Chirurgia non intermettendo però quando si possa tagliare, mostrare, e trattare anco le cose di essa Anatomia."

72. Ibid.

73. See *Venice, Città Excelentissima*, Appendix A.

74. Fabrici's salary in 1571 was 200 florins, and it doubled with his second appointment in 1577. In 1584, he became an ordinary professor of the first rank (*primo loco*), and, after assigning the lectures in surgery to Casseri, in 1589 he resumed his chair with an increased salary of 850 florins. In 1594, with an additional stipend, his salary was raised to 1100 florins. See *Embryological Treatises*, vol. 1, 7–8. On inflation, see Goldthwaite, *Economy of Renaissance Florence*.

75. *Statuti del comune di Padova* 33–34, 42–43.

76. Fabrici's relationship to the Venetian anatomist and scholar Paolo Sarpi (1552–1623) is well known. Less well known are the details of Fabrici's career as a surgeon. On his activities as a practitioner, see G. Favaro, "Girolamo Fabrici."

77. *Atti della nazione germanica*, 1585–1586, vol. 1, 210: "In qua sane diligentissime omnes non modo internas corporis partes, sed et externas, id est musculos atque vasa dissecando, insuperque praecipuas operationes chirurgicas monstrando optime sese gessit, nobisque omnibus satisfecit."

78. Ms. 677, f. 84r, AAUP: "Theatrum et cubiculi ad usum anatomiae agendae ex serenissimi Dominii benignitate de publico aerario eiusdem Dominii costructi sunt, ut in perpetuum preseverent ad ipsum usum destinati." This is cited by Gamba, "Il primo teatro anatomico," 158; and Rippa Bonati, "L'anatomia 'teatrale,'" 75–76. On the Massa family, see Palmer, "Nicolò Massa," esp. 407, n. 114.

79. *Atti della nazione germanica*, 1583–1584, vol. 1, 192: "Ut item deinceps sectiones publicae tam humanorum corporum quam brutorum quotannis ab Anatomico ordinario studiosis exhiberentur, quibus sane commodius et sine aliquo impedimento exercendis publicum et perpetuum theatrum in lectorio supremo sumptibus Dominii Veneti extrui secretarius curavit." Then, in vol. 1, 193: "23 Ianuarii, initium anatomiae publicae coepit h. 16 matutina, praesentibus aliquot doctoribus medicis, philosopho item Mercentario et metaphysico, ubi in novo illo theatro magnificis Venetis, una cum Rectore et professoribus primae sessiones, Consiliariis vero proximae fuerunt assignatae. Duravit haec diebus quatuordecim; satis diligenter administrata sunt omnia ab Anatomico, nisi quod nonnunquam a tumultuantibus nonnullis scholaribus Italis (ut mos eorum est) interturbatus, multa ipsius cogitationibus exciderint, quae alias maximo cum auditorum fructu discutienda in medium attulisset."

80. Ibid., 1588–1589, vol. 1, 267–68: "Ad ultimum Massarii etiam admoniti fuerunt, ut diligenter observarent, me loca illa, quae Consiliariis in anatomia destinata sunt, ab aliis occuparentur, atque illi vel abire vel de alio sibi loco prospicere cogerentur. XIIII die eiusdem mensis a magnifico huius urbis magistratu duo hominum propter furta et

maleficia perpetrata ad furcam damnati sunt; et quo minus anatomia publica impedire-tur aut differeretur, istorum corpora eadem statim die Universitati concessa fuere . . . a pluribus aliis vel Massariorum amicis vel popularibus obsiderentur, ut ita advenienti Domino Syndico apud Consiliarios nullus relinqueretur locus . . . Domino Syndico vix theatri limina transgresso, duo Siculorum, seu ut rectius dicam siccariorium, ex quat-uor illis Massariis obviam fiunt . . . ut hoc loco indignum explodere tentabant . . . Ubi Anatomicus intelligeret, rem hanc ad arma devenire, atque hunc ignem in immensum incendium quod postmodum quam difficilime consopiretur."

81. Tomasini, *Gymnasium Patavinum*, 77. See also Gamba, "Il primo teatro anatomico."

82. In the mid-seventeenth century, the Venetian Senate financed the construction of a permanent anatomical theater in San Giacomo dell'Orio for the *studio* of Venice. On the anatomical theater in Venice, see Molmenti, *La storia di Venezia*.

83. *Atti della nazione germanica*, 1586-1587, vol. 1, 222: "Concessae mense Septem-bris in Universitate feriae iam paulatim in finem decurrebant, ideoque publico program-mate ad aedem Cathedralem 2 Novembris cum scholaribus etiam professores a Rectore vocuntur. Ibi cum conventum esset, primo quidem eleganter et copiose de perfectione hominus et intellectus literis atque laboribus adipiscenda disseruit Reverendus quidam Monachus; eum in finem oratione directa, ut et hortaretur eos, quotquot essent sua per-fectionis studiosi, nullis parcerent laboribus, potissimum cum oblata esset in amplis-simo hoc literarum theatro occasio uberrima, et in laudem eorum digrederetur, qui Serenissimi Dominii munificentia hoc loco in Universitate Medica aluntur, non tantum Reverendi Theologi, verum etiam Medici experientissimi et sapientissimi Philosophi."

84. See Saibante, Vivarini, and Voghera, "Gli studenti dell'università," especially the statistical charts drawn from the university's extant matriculation records.

85. *Statuti del comune di Padova* 33-34, 42-43.

86. On the price of a cadaver, which was relative (and came to depend on how many cadavers were used in a demonstration), see chapter 4.

87. *Raccolta Minato*, n. 292 and n. 293, AAUP.

88. See chapter 1.

89. Heseler, *Andreas Vesalius' First Public Anatomy*, 221.

90. G. Favaro, "L'insegnamento anatomico," 125. On an uncertain date (after 1588), it was also noted that students were bringing arms into the theater.

91. On student violence, see Karras, *From Boys to Men*; and *Premodern Teenager*.

92. Marc-Antoine Muret, letter to Cardinal Sirleto (1584?), in *Renaissance Letters*, 43-44. Similarly, in 1545, Giuseppe Pallavicino lamented the weak response of the adminis-tration at the university of Bologna when he, while walking down the street, was "without reason struck a blow" from behind by one of his students. See Giuseppe Pallavicino, letter to Cardinal Morone (1545), in *Renaissance Letters*, 33-34.

93. *Atti della nazione germanica*, 1577, vol. 1, 123-24: "Nonas Novembris elegantis-sima oratione in studiorum laudem in ecclesia Cathedrali habita, ad diu intermissas lectiones reditum est. Professorum nomina, ut moris est, publice recitata, novique in defunctorum locum suffecti."

94. Ibid., 124: "Vixque ad VI Idus Novembris Doctores operas scholasticas potuerunt continuare, cum nonnulli, armorum potius quam literarum studio accensi, inter quos

multos exules et proscriptos fuisse constat, tumultus, qui in alterum usque mensem durarunt, excitare, docentesque impedire coeperunt."

95. In a study of the elegance and *copia*, or abundance, of Tuscan and Latin (1585), Aldo Manutio treated *impedimento* as a distraction from learning, as when friends come to visit and "impede my studies." Manutio, *Eleganze*, 162: "Io sono cosi spesso visitato dagli amici, che mi manca tempo per studiare: le visite de gli amici m'impediscono gli studi, mi sono d'impedimento, mi danno, porgono, apportono, recano, impedimento, talmentemi tengono occupato, che in tutto il giorno non ho hora libera per gli studi: gli amici col visitarmi spesso mi vietano l'attendere a gli studi, mi privano nel commercio de gli studi, mi tolgono ogni libertà, e ogni potestà di studiare, mi rimuovono da gli studi." The passage also appears in Latin in Manutio, *Eleganze*, 162–63.

96. *Statuta almae universitatis*, 33: "De forma servanda in universitate congregata: Statuimus per universitate congregata anteque aliquid proponatur Rector excludi faciat oès inhabiles ad vocem dandam: iubeatq, omnes scholaris sedere et pacifice audire rectorem proponentem, et alios loquentes super causa proposita. Nolentes sedere excludantur: nec quippiam proponi possit nisi exclusis contumacibus, et omnes sederint. Si universitas ad plures causas congregata est, nihil tractetur aut proponatur nisi prius expedito negocio principali: Quolibet autem partium seu res seiunctim et successive tractetur et expediatur, ita, ut duae res simul proponi et ballotari non possint. Nullum autem partitum rector aut quivis alius officialis ballotari permittat: nisi prius per notarium proclamari fecerit: si quis voluerit aliquid super negocio proposito loqui: accedat coram ipso rectore: et stans loquatur. Nullus autem loqui aut proponere quicquam posit nisi impetrata licentia vel a rectore: vel a consiliariis. Nec ullum verbum a quopiam fieri possit super aliqua re, nisi de qua tractanda per consiliarios fuerit obtentum, et volens aliquis quippiam dicere vel allegare super negotio tractando accedat ante rectorem ante cathedram eius. Ibiq; stans omnibus audientibus loquator modeste: et ordinate ut sibi suadendi aut disuadendi videbitur."

97. Ibid., 45: "De sedibus consiliariorum: In omnibus congregationibus nostrae universitatis adstent rectori, et sedeant in anteriori scamno: neque ullus scholaris in eo scamno in quo comsiliarii sedeni audeat sedere, sub poena sol. xx. in disputationibus vero publicis et processionibus per universitatem honorandis doctoribus immediate sequantur, in lectionibus vero audiendis similiter omnes scholares praecedant: neque in illis primis scamnis aliquis sedere possit exceptis consiliariis nisi solvat bidello duc. unum." In the early sixteenth century, Alessandro Benedetti described the hierarchical seating arrangement and the need to prevent disruptions in public dissections, reiterating what were most likely common practices and features of the demonstration. See Benedetti, *Historia corporis humani*, bk. 2, 1; and chapter 1.

98. *Atti della nazione germanica*, 1583, vol. 1, 192: "Ut denique ante administrationes anatomicas publice prius in ea ipsa doctrina iuventuti dilucidiores intelligentiae gratia aliquid praelegeretur, quo paratiores ad anatomicas contemplationes accedere possent."

99. Cunningham, *Anatomical Renaissance*, 191–211, and Carlino, *Books of the Body*, 126–28.

100. *Atti della nazione germanica*, 1584, vol. 1, 193: "[23 January 1584] initium anatomicae publicae coepit h. 16 matutina, praesentibus aliquot doctoribus medicis, philosopho

item Mercenario et metaphysico, ubi in novo illo theatro magnificis Venetis, una cum Rectore et professoribus primae sessiones, Consiliariis vero proximae fuerunt assignatae."

101. Ibid.: "Duravit haec diebus quatordecim; satis diligenter administrata sunt omnia ab Anatomico, nisi quod nonnunquam a tumultuantibus nonnullis scholaribus Italis (ut mos eorum est) interturbatus, multa ipsius cogitationibus exciderint, quae alias maximo cum auditorum fructu discutienda in medium attulisset. Crebrae praeterea obiectiones ibi a professoribus motae et utrinque agitatae fuerunt. Tandem sectione capitis humani et ovis foetae viventis absoluta, finem imposuit."

102. The transalpine students describe him as "Paulo Galeoti viro Nationis nostrae studiosissimo meritissimo" in *Atti della nazione germanica*, 1591, vol. 1, 303.

103. *Atti della nazione germanica*, 1584–1585, vol. 1, 202: "Dominus Hieronymus Fabritius de Aquapendente aggressus est anatomiam publice exhibendam, eamque in undecimum usque diem protraxit, omissa hoc anno dissectione vivi corporis, quam tamen saepius pollicitus fuerat."

104. Ibid., 210: "Cuius initium atque progressus pro more clarissimi Anatomici ut satis fuit gratus ac laudabilis, cum praesertim in viscerum monstrationibus magna uteretur diligentia proferendo iecur, lienem, renes, nec non musculum carne remota, itemque uterum gravidae mulieris una cum placenta, aliaque non iniucunda visu."

105. Fabrici eventually published two studies on generation. See his "De formatione ovi et pulli" and "De formato foetu" in *Embryological Treatise*.

106. *Atti della nazione germanica*, 1585–1586, vol. 1, 210: "In qua sane diligentissime omnes non modo internas corporis partes, sed et externas, id est musculos atque vasa dissecando, insuperque praecipuas operationes chirurgicas monstrando optime sese gessit, nobisque omnibus satisfecit."

107. Ibid., 211: "Quocirca nos hoc exercitium e re omnium nostrum fore rati (cum praesertim partium usus diligentissime sit explicaturus, accommodando omnino methodo Galeni a praeclarissimo Domino Capivaccio multis partibus applicitis) hoc vobis significare hisce voluimus. Rogamus itaque universos et singulos, ut benevolentiam et labores huius clarissimi viri, propriamque utilitatem considerando, frequentes huic interesse velint."

108. Ibid., 1586–1587, vol. 1, 225–27. Also see n. 101.

109. Ibid., 225: "Absolvit interim anatomiam suam Galeotus, quam multis diebus diligenter satis et in excitante eum auditorum frequentia hactenus administraverat. Cumque paulo post sponte Italos nonnullos gratitudinis ergo in theatri anatomici exstructionem contulisse symbola praedicaret, a nostris idem sese sperare haud obscure significabat; potissimum cum sciret nostram prae reliquis Nationem anatomicis ut plurimum delectari studiis, haecque res omnis non in suam sed auditorum suorum utilitatem esset redundatura."

110. Ibid.: "Excellentissimi Anatomici tam lauta fuisse promissa dixi, ut nihil praeter anatomica exercitia totam hyemem expectaverimus."

111. Ibid., 1585–1586, vol. 1, 210.

112. Ibid., 1586–1587, vol. 1, 225–26: "Facto autem principio, post declarata externa corporis lineamenta cadaveris ratione protinus ad oculum aggressus Anatomicus, de

eius actione et fabrica ita plane ita erudite multis disseruit lectionibus, ut virum sese gesserit anatomicarum rerum et opticae peritissimum."

113. On Galeotto's 1588 demonstration, see n. 131. For his 1591 demonstration, see *Atti della nazione germanica*, 1591, vol. 1, 303: "Galeotus vero summa diligentia omnium partium cognomen ut quisquam alias omnium iudicio perfectissimam exacte proposuit. Conveniens certe fuisset et gratis auditoribus dignum, Galeoto pro immensa et liberali sua industria si nobis nullus modus reprehensione et invidia carens occurrebat alius, is constituebatur, uti verbis saltem Nationis nostrae nomine ipsi gratiae agerentur, merita ista ipsius perpetuae nostrorum memoriae commendarentur, sique alia de nostra Natione in ipsum proficisci possent gratitudinis munia, ea prompta parataque offerrentur omnia."

114. *Atti della nazione germanica*, 1587, vol. 1, 244: "Circa principium Studii, mandato Clarissimorum huius urbis Rectorum sub poena banni, ne quis in posterum contra consensum Universitatis in scholis alicuius arma erigat, neve quis ante publicum finitam, privatam aliquam anatomiam instituat, fuit interdictum. Ex hoc igitur privilegio quod in fraudem scholarium ad obvelandam solitam suam negligentiam a Reformatoribus fucosis argumentis consecutus fuerat, occasionem capiens publicus Anatomicus, theatrum illud anatomicum Eccellentissimi Domini Pauli Galeoti scholarium impensis extructum destruxit, sed id procul dubio si citius rescivissemus, aggredi non fuisset ausus, quod post malis artibus consecutum privilegium, libidinose exequi non dubitavit."

115. Ibid., 244–45: "Postquam autem post paucos dies praedictus Magnificus Vicerector Patavium esset reversus, et a nobis ad anatomiam urgendam admonitus, sine cunctatione eos qui Massarii seu anatomistae fieri vellent, schaeda quadam publica ad se vocavit. Anatomicus interim Venetiis agens, de nulla re minus quam de anatomia cogitat, sperans cum ante ipsius publicam, nulla privata institui a quovis posset, se (quidem et palam postea profiteri non erubuit) vel nullam omnino, vel saltem suo more superficialem aliquam habere, reliquis autem commodum sectionibus tempus praeripere hoc pacto posse."

116. Ibid., 245: "Cum itaque, ut aequum erat, rem hanc indigne ferret, pertaesus tandem nos cum aliis Germanis secum Anatomicum iussit accedere, quo si minus sua is authoritate moveretur, saltem nostris precibus locum relinqueret. Atque ita 5 vel 6 Germanis comitati, una cum Magnifico Vice Rectore aedes Aquapendentis accedimus, ipsum absentem per horam fere expectamus, rogamus, supplicamus ut tantopere desiderata, tum ab omnibus aliis tum vero potissime Germanis (quorum multi huius solius exercitii gratia appulissent) anatomiae faceret principium."

117. Ibid.: "Tum ille multa de sua in Germanos affectione (quos non nisi blandis verbis decipere consueverat) praefatus, de anatomia omnino hoc anno fienda spem nobis facit; verum nos, qui ex antecessorum nostrorum Actis, spem pretio non emere didiceramus, hisce non cnotenti, de tempore et aliis circumstantiis quaesivimus, ad quae cum modo hoc modo aliud responderet, atque etiam de anatomia hoc anno celebranda addubitare videretur, ut cathegorice tandem, quid nobis de anatomia sperandum esset, responderet, aliquoties ipsum hortati sumus."

118. Ibid., 246: "Alia itaque universitas habetur, et ex malis quatuor, duo meliores eliguntur (quidem id ex actis Universitatis super hac re in aerario nostro repositis latius videre licet)."

119. Ibid.: "Ille vero in has angustias reductus, bilem aliquantulum concipere coepit, quod cum parum nos curare cerneret, respondit: Vos tantopere anatomiam urgetis, cum tamen neque de Massariis, neque cadaveribus vobis sit prospectum."

120. Ibid., 245: "In quibus vero cum moram non esse ullam illi monstraremus, sui officii esse anatomistas Universitati proponere significavit, quod licet manifesto Universitatis statutis utpote 28 lib. 2 contradicere videbatur, tamen ne nobis alias turbas moveret, et ne tempus anatomiae debitum rixis et contentionibus perderemus, potius aliquid de nostro iure remittere, quam utilitati scholarium obesse volentes, et id quod falso sibi praesumebat pro ea vice concedentes, ipsi ut quatuor, ex quibus Universitas duos eligeret, nobis intra quatriduum proponeret scholares ad hoc ministerium idoneos permisimus."

121. Ibid., 246: "atque ita post aliud quatriduum quatuor adolescentulos scholam Iesuitarum frequentantes, non modo huius exercitii omnino imperitos, sed et omnis medicinae penitus ignaros Universitati proposuit."

122. Ibid., 246–247.

123. Ibid., 234: "Excellentissimus igitur Dominus Hieronymus Capicavius, post absolutam de capitis affectibus exquisitissimam doctrinam, adeo doctum et perfectum de omnibus oculorum malis tratatum nobis praelegit, suasque lectiones ad 11.m. Iulii usque diem produxit, ut maximam ipsius diligentiam cum summa eruditione semper coniunctam esse non obscure nobis omnibus ostenderet."

124. Ibid.: "Eccellentissimus Dominus Hieronymus Mercurialis perfecto tractatu suo de vitiis supremi ventris, thoracis quoque affectus usque ad cordis mala absoluit, suisque auditoribus 30 Iunii valedicens . . . quos tantopere amavi, et quos amo, multis cum lachrymis et non sine dolore sim imo vero ditissimum omnium scientiarum mercatum." Mercuriale's place was soon taken up by Alessandro Massaria (1510–1598).

125. *Atti della nazione germanica*, 1587, vol. 1, 235: "Quam diligens fuerit Excellentissimus Dominus Albertinus Bottonus in suis quotidianis discursibus, omnibus fere nostris notum est. Nam singulis diebus nos ad novum patientem deduxit, deque illius affectu, causis, signis, indicationibus, ac prognosticis doctissime disseruit, proponendo non solum materias remediorum accomodatissimas, verum etiam semper aliqua quae in secretis habebat adiungendo, ut huius viri singularis benevolentia et humanitas, qua multis nunc annis nostram Nationem prosectus est, hoc quoque et praecendenti anno plurimus perspectissima fuerit."

126. Ibid.: "Etsi enim theoria quam omnes fere ex patria nobiscum adducimus, nos aliquo modo ab illius lectionibus deterrere possit; tamen illius in praxi mira felicitas et accuratissima curandi ratio, nos iterum invitare solet."

127. Ibid.

128. See G. Favaro, "L'insegnamento anatomico," 116.

129. Ibid.

130. *Atti della nazione germanica*, 1589, vol. 1, 271.

131. Ibid., 1588–1589, vol. 1, 248: "His itaque peractis, et publica anatome tandem ad finem perducta, Eccellentissimus vir Dominus Paulus Galeoti ne quid a pristina sua de

scholaribus bene merendi voluntate recessisse videretur (quam vis iniuriis publici Anatomici theatrum ipsius esset destructum, ipse variis modis ab eodem fuerit lacessitus et in minus commoda carnisprivii tempora reiectus), in officina *al Corallo* per integras tres septimanas luculentissimam et absolutissimam habuit anatomiam, qua non solum mira facilitate ac pulchra methodo, modum secandi corpora, partium omnium structuram ac earundem actiones et usus evidentissime monstravit, sed etiam venarum, arteriarum et nervorum in universo corpore ductus et propagines, non sine maxima omnium nostrorum iucunditate, sine ullis nostris impensis (quamvis ad eas ut aequum erat a spectatoribus exigendas a nobis hortaretur) nobis ostendit."

132. Galeotto's pedagogical style may be characterized as clear, concise, comprehensive, and inspirational. The transalpine student recorded his "merits" and the students' desire to commend him and advance his status at the university. See *Atti della nazione germanica*, 1588–1589, vol. 1, 248–49: "cum igitur tanta, non solum in totam Universitatem sed et potissimum in nostram Nationem, huius viri sint merita, isque pro hisce suis laboribus et vigiliis nullum aliud praemium reportarit, sed Nationi praesertim nostrae quam commendatissimus esse cupiat. Quocirca successores nostros etiam atque etiam hortamur et amice rogamus, ut si aliqua gratum nostrum animum huic viro declarandi sese obtulerit occasio, nullis parcant laboribus, sed omnes nervos eo dirigant, quo is non solum contra invidiam et malevolentiam aliorum defendatur, sed etiam, quibus modis id fieri poterit, commendetur, ornetur et promoveatur."

133. *Atti della nazione germanica*, 1588–1589, vol. 1, 267–68.

134. Ibid., 1590, vol. 1, 290: "Ad XII Kal. Aprilis privatae anatomiae tantopere desyderatae administrationem aggreditur Excellentissimus Paullus Galleottus . . . Quamquam autem tempus anni maturandam suadebat, tamen succincta brevitate omnia complexus, parerga praecidens, et fragrantes rerum succos libans, nihil praeteribat quod ad rem faceret; immo nervorum, arteriarum et venarum perfectam ostensionem nobis exhibuit."

135. See notes throughout 1589–1590 in *Atti della nazione germanica*. In 1589–1590, and again in the following year, Fabrici's demonstration was suspended; there were student protests and, in 1592–1593, amid diverse disruptions, the students found the demonstration laborious.

136. The letter is transcribed in Sterzi, "Guilio Casserio," esp. 74–75 [pt. 2].

137. *Atti della nazione germanica*, 1590, vol. 1, 286: "Iam ossium capitis enarrationi et descriptioni duos impendit menses: ad musculos devolutus tres absolvit singulis musculis singulas horas tribuens. Tot autem enumerantur musculi, ut hac via incedens biennium non sufficiat. Quando igitur de visceribus? Accedit quod confuse et tumultuarie omnia pertractantur: iam branchium detruncatum attulit, post multos dies pedem allaturus, quando aliquis ex horum inspectione seriem et connexum totius discere posset, non video." Cited by Cunningham, "Fabricius and the 'Aristotle Project,'" 199 n. 7.

138. *Atti della nazione germanica*, 1590, vol. 1, 288: "Quid, inquit, privatam desideratis, nunquid placuit publica? Exactissimam, respondemus, nobis dedit Aquapendens; verum ut illa quae fusius et licentiori sermonis genere tradidit Anatomicus, nobis iam quasi in synopsi ob oculos proposita memoriae mandemus, de privata agimus."

139. Ibid., 1590–91, vol. 1, 301: "Appropinquantibus iam brumae frigoribus anatomiae administrandae opportunis, Anatomicus exactissimam et 4 menses duraturum anatomen pollicetur. Cum vero nundum se pararet ad anatomen, Dominus Ioannes Hertelius alias Syndicus Universitatis et Marcus Antonius Ponderanus Cretensis, pro Massariis ab Universitate X Decembris congregata electi, eum edeunt, operas suas in inveniendis cadaveribus et procurandis ad anatomen necessariis offerunt, et ad labores inchoandos hortantur."

140. Ibid., 303: "Impetrata itaque licentia a Praetore, per Universitatem 22 Ianuarii eligebatur Excellentissimus Paulus Galeotus, ob plures ingeniose exhibitas anatomias iam ante celebris, eique munus faciendae extraordinariae anatomiae imponebatur, et cume haec publicae anatomiae vicem supplere deberet, ut sine confusione omnia gererentur."

141. Ibid.: "Galeotus vero summa diligentia omnium partium cognomen ut quisquam alias omnium iudicio perfectissimam exacte proposuit."

142. Ibid., 1592–1593, vol. 2, 32: "Quoniam vero tunc temporis Theatrum Anatomicum erat destructum, rumorque spargebatur fore ut hoc anno minime restauraretur, ego ut multorum ex nostris petitioni satisfacerem, nullamque occasionem pro anatomia publica promoverda negligerem, accessi Anatomicum eumque ad publicam sectionem sese ut preparet sollicitavi. Qui, dummodo theatrum reficeretur, nihil se aliud optare respondit, multis verbis suam in ea administranda diligentiam promittens."

143. Ibid.: "Iterea autem temporis publice methodum suam anatomicam ante biennium praelectam repetiit, privatim vero domi suae anatomiam foetus celebravit. Qua absoluta, oculi dissectionem instituit, publice de visione disserens."

144. See Rupp, "Matters of Life and Death"; and Blancken, *Catalogue*.

145. *Atti della nazione germanica*, 1594, vol. 2, 56: "Et ecce Aquapendens aurearum suarum pollicitationum immemor, memor vero pristinae et iam per annos aliquot obduratae calliditatis, novas offucias quibus varie Universitatem agitabat comminiscitur. Quippe nihil egregie, nihil utiliter, nihil frugifere se spectatoribus ac auditoribus condocefacere aut demonstrare posse saepe revomuit."

146. The phrase comes from Jacopo Tomasini's description of the anatomical demonstrations inside the second theater. See Tomasini, *Gymnasium patavinum*.

CHAPTER 3: CIVIC AND CIVIL ANATOMIES

1. *Atti della nazione germanica*, 1588–1589, vol. 1, 266 and following. The students who wrote this were Urbano Zussnero Carniolano and Eberhardo Vorstio Geldro Ruramondano, president and vice-president for the transalpine nation.

2. *Atti della nazione germanica*, 1588–1589, vol. 1, 267: "Ad ultimum Massarii etiam admoniti fuerunt, ut diligenter observarent, me loca illa, quae Consiliariis in anatomia destinata sunt, ab aliis occuparentur, atque illi vel abire vel de alio sibi loco prospicere cogerentur. XIIII die eiusdem mensis a magnifico huius urbis magistratu duo hominum propter furta et maleficia perpetrata ad furcam damnati sunt; et quo minus anatomia publica impediretur aut differeretur, istorum corpora eadem statim die Universitati concessa fuere."

3. Ibid.: "Anatomicus audita iam hora lectionis initium fecisset, loca illa, quae tantum pro Consiliariis retinenda essent, a pluribus aliis vel Massariorum amicis vel popularibus obsiderentur, ut ita advenienti Domino Syndico apud Consiliarios nullus relinqueretur locus." On class categories, see Brian Pullan, *Rich and Poor*.

4. *Atti della nazione germanica*, 1588–1589, vol. 1, 267–68: "Syndicus autem Universitatis atque amplissimi sui offitii ratione statim ab initio anatomiae apud Doctores et Professores in theatro locum obtinere potuerat; ne se ipsum tamen ingerere videretur, Dominis Consiliariis se adiungere maluit. Cum igitur se aliquantulum tardius accessisse, atque loca, facto lectionis initio, replete videret, potius sibi abeundum quam strepitum aliquem excitandum aut legentem Anatomicum interturbandum esse, cogitavit."

5. *History of the University in Europe*, vol. 2, 173.

6. *Atti della nazione germanica*, 1588–1589, vol. 1, 268: "Sed bidellus Universitatis, qui ad theatrum expectare atque advenientes Professores ad assignatas sessiones deducere solebat, videns in theatro quaedam Professorum loca vacua superesse, Domino Syndico locum dari Massarios iubet . . . Etenim Domino Syndico vix theatri limina transgresso, duo Siculorum, seu ut rectius dicam siccariorium, ex quatuor illis Massariis obviam fiunt, qui non solum posthabita omni Universitatis et praesentium Excellentissimorum virorum authoritate, sibilis et ignominiosis verbis ipsum ut hoc loco indignum explodere tentabant, sed etiam alter ipsorum exuto omnis humanitatis offitio, stricto pugione impetere nequaquam subverebatur."

7. Ibid.

8. Ibid.: "Contra tamen Dominus Sapiens, ut vere sapientem atque fortem virum decet, sufficienter refutatis atque convitiis in adversaries retorsis, strenue se opposuit, neque latum quidem unguem ab eo quem occuparat loco dimoveri passus fuit. Ubi Anatomicus intelligeret, rem hanc ad arma devenire, atque hunc ignem in immensum incendium quod postmodum quam difficillime consopiretur, excrescere et hinc propter vicinum malum ad se quoque aliquid mali pervenire posse, confestim Siculis serio interdixit, ne quid ulterius contra Syndicum, cui maxime hic locus competeret, susciperent."

9. See Gamba, "Il primo teatro anatomico."

10. *Atti della nazione germanica*, 1588–1589, vol. 1, 268: "Interim Anatomico in lectione pergente, quamplurimi Germanorum atque nostrarum partium Italorum convocantur, qui partim in collegio, partim prope collegium bene armati expectabant, ut si quid adhuc adversarii violenter molituri essent, resistere illis possent."

11. This link with Carnival derives from the particular case of Bologna and the anatomical practices and procedures at its university, as argued by Giovanna Ferrari in "Public Anatomy Lessons."

12. On student violence, see Karras, *From Boys to Men*; and Grendler, *Universities*, 500–505. For evidence of the violence between students, see *Relazioni dei rettori veneti*, 149; and Niccolò Rossi, *Annali di Padova*, Library of the Civic Museum, Padua (hereafter cited as MCP).

13. *Atti della nazione germanica*, 1588–1589, vol. 1, 268: "Sed finita lectione, Massarii veniam a Domino Sapiente [Syndic] deprecantes, intercedente Anatomico, pacem cum illo inierunt."

14. The transalpine students noted that the anatomist Paolo Galeotto was forced to wait until the period of Lent to offer a private dissection. Galeotto was not a chaired professor, so his lessons would naturally follow those of Fabrici. In the preceding years, it was typical for both Fabrici and Galeotto to complete their lessons before Carnival began. The delay of Galeotto's private dissection until Lent suggests that Fabrici's demonstration, as well as the ensuing conflict, took place during Carnival. See *Atti della nazione germanica*, 1589, vol. 1, 270.

15. Mikhail Bakhtin organized his concept of the carnivalesque around play, the motif of the world turned upside down, and grotesque realism. On the last, Bakhtin writes: "Contrary to modern canons, the grotesque body is not separated from the rest of the world. It is not a closed, completed unit; it is unfinished, outgrows itself, transgresses its own limits. The stress is laid on those parts of the body that are open to the outside world . . . the apertures or convexities . . . the open mouth, the genital organs, the breasts, the phallus, the potbelly, the nose. The body discloses its essence as a principle of growth which exceeds its own limits only in copulation, pregnancy, childbirth, the throes of death, eating, drinking, or defecation. This is the ever unfinished, ever creating body" (Bakhtin, *Rabelais and His World*, 26). Subsequently, scholars have historicized and qualified different parts of Bakhtin's analysis. See Burke, *Popular Culture*, 178–243, on the reform of Carnival practices by Protestants and Catholics; Muir, *Civic Ritual*, 13–64, on the mutual influence of myth, ritual, and social action in Venice; and Ciappelli, *Carnevale e Quaresima*, for a local analysis of the structural links between Carnival and Lent and the ways in which Lent, with its emphasis on private life and privation as well as on liturgical and devotional practice, shaped the experience of Carnival in late medieval Florence.

16. On the reforms of Carnival practice in the 1520s in Venice, see Muir, *Civic Ritual*, 156–81. For an example involving nuns, see Ruggiero, *Boundaries of Eros*, 78–79.

17. See chapter 4. In addition, see Park, "Criminal and the Saintly," esp. 12; and Carlino, *Books of the Body*, 118–19.

18. The danger of the Counter-Reformation for foreign students developed more slowly in Padua than in Bologna. For this reason, Padua continued to attract foreign students. See Kagan, "Universities in Italy"; and Saibante, Vivarini, and Voghera, "Gli studenti dell'università." There was, however, trouble in Padua. See Brugi, *Gli scolari dello studio*.

19. On (Italian) Renaissance pedagogy, see Eugenio Garin, *Il pensiero pedagogico*. On comportment manuals, see Patrizi, *La trattatistica educativa*; and Botteri, *Galateo e galatei*.

20. For example, see Plautus, *Aulularia* I.1.18; and Ridgeway, *Origin of Metallic Currency*, 354. On the biblical sense, Edward Campion cites "quia nos ne latum quidem unguem discedimus a verbo Dei." See Campion, *Ten Reasons*, 228. I thank Evan Hayes for querying this strange construction and for these sources.

21. In updating Norbert Elias's study, *The Civilizing Process*, which focuses on courtly manners and the formation of the bourgeois subject, Steven Shapin has studied civility in the context of the Royal Society in England and argued that these manners and codes of civility cultivated disinterestedness, which gradually began to grant their endeavors a degree of certainty (when it inhered in their knowledge claims). Whereas Shapin sees this experimental context and its codes as alternatives to university learning and as espe-

cially English (dependent on anxieties around class mobility among the landed gentry), in this chapter I show that the genealogy of that civility may be found at the university, in the pedagogical programs of Renaissance humanists, including medical humanists, and in the concerns for obedient, silent, and attentive students. See Shapin, *Social History of Truth*, 9–14, 42–74. In contrast, Lorraine Daston has shown that similar codes of civility and sociability pervaded early scientific societies on the Continent as well. See Daston, "Baconian Facts," 37–45.

22. Abriano, *Annali di padova*, 135v, MCP: "Furono rinfatti il dormitorio, et altri luoghi abbruciati dal fuoco dii lor S. Giustitià. Et nelle scole del Bue fu atto il luoco da fer l'Anotomia."

23. Its streamlined effect has been described as similar to the celebrated Venetian tradition of naval carpentry. The architect of the theater remains unknown. Critics have suggested Fabrici, as well as Dario Varotari, a friend of Fabrici who, in addition to being a painter, was also the architect of Fabrici's villa. See Dal Piaz, "L'orto botanico," 69; Rippa Bonati, "Le tradizioni"; and Cagnoni, "I teatri anatomici" (unpublished).

24. Tomasini, *Gymnasium patavinum*, xxxi, "De anatomia et anatomicus"; Gamba, "Il primo teatro anatomico," 158, 160 n. 15; Rippa Bonati, "L'anatomia 'teatrale.'"

25. Underwood, "Early Teaching of Anatomy," esp. 17.

26. Underwood, "Early Teaching of Anatomy," 16. Underwood imagined himself turning around in each of the tiers, pressing against his imaginary neighbor, and generated the number 244 for the quantity of potential spectators. However, he suggests that this number was probably lower, due to the voluminous robes that were a part of sixteenth-century garments.

27. Malfatti, *Descrizione particolare*, 54.

28. Tomasini, *Gymnasium patavinum*, xxxi, "De anatomia et anatomicus," 78: "Adversa altera conclave aperit, in quo praesentibus variarum nationum studiosis, quae publicae demonstranda sunt, apertis cadaveribus exscindi ac praeparari consueverunt. Huius fores et parietas Anatomicorum insignia exornant, grata Universitate, tum ingenio docentium, tum laboribus naturae arcana scrutantium, honorem hunc exhibente. Ex eo in spatium aliud porta ducit, in cuius secretiore angulo cadavera, eorumque partes dissectae asservantur. In eodem arrae sunt et conditoria, vestimentis, sceletis, instrumentisque Anatomicorum custodiendis aptata."

29. Underwood also describes the alcoves and stonework in this lower room. See Underwood, "Early Teaching of Anatomy," 18. The fact that stone is colder and may better preserve cadavers may have influenced the construction of the anatomical theater in Venice, which was built with stone. On that construction, see Molmenti, *La storia di Venezia*; Rizzi, *Il teatro anatomico*; Pasinetti, "Il ponte dell'anatomia"; Cecchetti, *Per la storia della medicina*; and Bernardi, *Prospetto storico-critico*. For an oration celebrating its opening, see Grandi, *Orazione*.

30. Underwood, "Early Teaching of Anatomy," 18.

31. Andrews, *Scripts and Scenarios*, 31–63, and for Padua and Venice, 121–68.

32. *Statuta almae universitatis*, bk. 1, "De forma servanda in universitate congregata," 33, and "De sedibus consiliariorum," 45. In addition, see *History of the University*, vol. 1, 195–230, esp. 205.

33. The private venues of anatomy are the subject of chapter 5.

34. The full episode is discussed in chapter 4. See *Atti della nazione germanica*, 1595, vol. 2, 58–59: "Hanc petulantiam ac execrandum fastum Anatomistae, qui indefessa diligentia ut rite et cum laude publicae sectionis administrationes in novo illo theatro procederent laborabant, aegerrime tulere." In the previous year, as Giuseppe Sterzi notes, the transalpine students had thought it necessary to elect a *consigliere anatomico*, a kind of preparer, to help Fabrici with the labors of dissection. See Sterzi, "Guilio Casserio," esp. 213, 232–35 [pt. 1].

35. *Atti della nazione germanica*, 1600–1601, vol. 2, 180: "11. d. Novembris Celeberrimus noster Anatomicus, Dominus Hieronymus Fabritius ab Aquapendente, absoluta prius in auditorio methodo anatomica generali, deinceps in Theatro anatomias foetuum, et quidem equini et agnelli aggressus est. 20. d. Novembris, convocata Universitate Magnifica, electi sunt Anatomistae, quos vulgo Massarios vocant, qui Excellentissimo Anatomico in dissecandis et procurandis cadaveribus auxiliares manus ferre solent. Proposui ex nostra Natione Doctissimum Dominum Hectorem Selanofium Hessum, qui cum duobus Italis huic muneri praefectus est."

36. Ibid., 1588–1589, vol. 1, 267: "Ad XVI diem huius, Aquapendens hora XVI dissectionem structurae humani corporis aggressus est. Quamvis autem Aquapendens in huius anatomes administratione XIIX dies integros consumpserit, atque paulo diligentionem atque superioribus annis se gesserit, aureis tamen suis pollicitationibus minime satisfecisse videbatur."

37. See Park, "Organic Soul." With regard to Renaissance anatomy, see French, *William Harvey's Natural Philosophy*, 10, 66. With regard to Fabrici's program, see Cunningham, *Anatomical Renaissance*, 167–87.

38. Daston, "Baconian Facts," 37–63. Andrea Carlino notes that Aristotle's works made observations more significant through the use of analogies between animal and human physiology. See Carlino, *Books of the Body*, 133.

39. Peter Dear has shown that by the 1650s, the category of experience signified the rare and the particular rather than the common. See Dear, "Jesuit Mathematical Science."

40. "De formatione ovi et pulli," in *Embryological Treatises*, 142.

41. Tomasini, *Gymnasium patavinum*, 78.

42. On the distinction between secrets of nature and marvels (such as monsters), see Daston and Park, *Wonders*, 173–214. For the tradition of secrets related to natural magic—where *secretae* derive their meaning as something hidden underneath nature's surface, tinged with impiety and a curiosity about the occult—and their presence in the print cultures of the Renaissance, see Eamon, *Science*, 194–266, esp. 194–203. On gendered aspects of this history, see Park, *Secrets of Women*, 77–120; and Green, "From 'Diseases of Women.'"

43. On the tradition of the secrets of nature, see Eamon, *Science*.

44. *Atti della nazione germanica*, 1595, vol. 2, 58: "Confluxerat eo tota quasi civitas, et extremae etiam farinae homines tanquam ad forum cupedinis: subsellia occuparunt hebraei, sedentarii magistri, sartores, calceolarii, solearii, carnarii, salsamentarii et his inferiores baiuti et corbuli illi, adeo ut in dubium relinqueres plus ne collegii scholares

anatomici sectioni ac dexteritati attenderent, an haintia huiusmodo homuncionum ora aspicerent."

45. See Bylebyl, "School of Padua." On the ways in which seventeenth-century Venice turned inward, losing much of its cosmopolitan perspective, see Burke, "Early Modern Venice"; and, more recently, de Vivo, *Information and Communication*.

46. Saibante, Vivarni, and Voghera, "Gli studenti dell'università," esp. the matriculation graphs.

47. These nations were geographically designated and, in Padua, included cisalpine and transalpine areas. On the student nations in Bologna and Padua, see Kibre, *Nations*, 3–64, 116–22. On the role of foreign students in Italian Renaissance universities, see the following and their bibliographies: Kagan, "Universities in Italy"; and *History of the University*, vol. 2. On foreigners in Padua, see Fedalto, "Stranieri a Venezia," esp. 271–78.

48. *Atti della nazione germanica*, 1597, vol. 2, 110: "Certo si absque Ultramontanis esset, Theatrum saepe vacuum quin reperiretur, vel Excellentissimo Viro teste affirmare ausim."

49. The Senate's decree accompanies the inaugural oration that Cesare Cremonini gave on 20 December 1591, in which he lambasted the Jesuits and the divisiveness they sowed in the student body. See *Padova, Studio. Documenti 1467–1625*, Library of the Correr Museum: "Ricerca la dignità è servitio del studio nostro di padoa che li dottori forestieri condotti da queste cons. per erudir li scolari in queste scienze che tal sono necessarie per ben regere ogni Republica, essendo queste persone conspicue di gran valore, è stima appreso il Mondo tutto, sino honorati et ben trattati da ogni anno et spetialmente da gli altri dottori della Citta accioche con tal ragioneval modo aletati si rendino sia pronti ad incontrar le occasione di venir à questo servitio."

50. Brugi, *Gli scolari dello studio*, 86–91. Before they made requests, transalpine students would often note that they had looked back in their records and found certain professors sympathetic to their past causes and, most likely, promoters for their current ones. Cesare Cremonini and Giambattista Da Monte were described as such.

51. *Atti della nazione germanica*, 1595, vol. 2, 57: "Sed volo ut sit libera, omnes ut videant absolutissimam; et oportet ut Universitas Venetiis scribat ut ex pecuniis Studii sumtus habeat, et ego etiam intercedam meis literis; et concedo ut eligatis secundum statuta Massarios quibus singulis septimanis dabo florenos quatuor." On free admission, see also Riccoboni, *De gymnasio patavino*, xviii, "De ingressu in theatrum ad spectandam anatomen et honore anatomici."

52. *Statuta almae universitatis*, bk. 2, "De anothomia singulis annis fienda," 27.

53. *Senato Terra*, Registro 66, ASV: "L'andera parte, che per l'avvenir l'Anatomia nel predetto studio nostro di Padoa sia libera, si che cadauno possa entrar nel theatro à vederla senza pagar cosa alcuna, et per le spese, che occorrono farsi siano delli danari delli danari [sic] della cassa del studio ogn'anno che si farà detta Anatomia Ducati XXV. da esser dati a parte a parte, secondo che farà bisogno, il qual bisogno sia conosciuto da lettor di essa alli Massari over Anatomisti da esser eletti dall'Università al principio del studio per provinar le cose necessarie d'essa Anatomia non potendosi dar più di fiorini quattro alla settimana; per quelle settimane però, che l'Anatomico taglierà, et amministrerà le parti del corpo, et non quato solamente leggerà, et detti sfiorini quattro hanno

contati in mano dell'Anatomico, che taglierà, et administrerà et per lui siano dati a detti Massari secondo che conoscerà esser bisogno."

54. The specificity of university statutes, Andrea Carlino explains, sought to limit the potential for disorder: "scuffles that might break out over the unsuitable choice of a cadaver," disturbances that "might be provoked by relatives and friends who protested the profanation of the remains of their loved one, by spectators at the dissections, by barbers and surgeons over the exorbitant entrance fees, or by controversies that could arise from the assignment of places normally distributed on the basis of the position, seniority, and dignity of those present." See Carlino, *Books of the Body*, 78–79, 83–84.

55. *Senato Terra*, Registro 66, ASV: "Perche l'Anatomia tanto necessaria alla Medicina, et cognitione tanto degna d'ogni studioso fosse detta, et tagliata nel studio nostro di Padoa con quella dignità che si ricevea à cosifatto studio, et con quel frutto, che si deve aspettar da cosi importante lettura et materia che si può dir delle più principali delle arti et medicina si è fabricato in quelle scuole nostre il theatro per farla in esso stabile, et honoratissimo, resta nondimenso à farsi provisione, esse non sia disturbata, come per l'aversoro [*sic*] si è fatto ogn'anno con maleficio de'scolari, et con non poca indegnità nostra pero."

56. Tomasini, *Gymnasium patavinum*, 78.

57. Ibid.

58. On the use of processions in Venice to represent symbolic order, see Muir, *Civic Ritual*, 185–211. On music used during such occasions, see Rosand, "Music."

59. Tomasini, *Gymnasium patavinum*, 79: "Solemnis enim celebratur concursus et pomposa praeparatio omnium. Coronantur Lauro Gymnasii limina cum stemmatibus illustrissimorum Rectorum Civitatis, Rectoris Universitatis; et Anatomici, musica interdum Doctoris orationi praemissa."

60. *Atti della nazione germanica*, 1597, vol. 2, 111: "A.d. 12 Xbris ad exhilarandos anatomiae spectatores recreandosque ex tristi aspectu animos, ex vetusta consuetudine (quae tamen superioribus aliquot annis proximis interrupta) fidicines ab Anatomistis conducti et in Theatrum reducti fuere, procurante hanc rem sedulo D. Placotomo; aderantque musici isti etiam sequentibus diebus quamplurimis, sumtus certe qui illis irrogantur minime poenitendi, si quidem dum ipsis attendunt et auscultant spectatores, ab omni tumultu et calcitratione supersedere solent, cuius tranquillitatis gratia Theatrum anatomicum aliquot annos diutius inconcussum durare poterit." The passage is also cited by Gamba, "Il primo teatro anatomico," 160.

61. Tomasini explains that students took part in musical exercises as well as dancing lessons, military exercises, and physical ones. See Tomasini, *Gymnasium patavinum*, xxxvii, "De aliis exercitiis gymnasticis," 133. For Francesco Portenari, see Brunelli, "Francesco Portenari"; and Brunelli, "Due accademie padovane."

62. Tomasini, *Gymnasium patavinum*, 79.

63. Disputations were verbal exercises that engaged students, lecturers, and professors. They followed the structure of narrow debates, including point, counterpoint, and resolution. On the medieval and Renaissance traditions of university disputation, see Costello, *Scholastic Curriculum*; Grendler, *Universities*, 152–57; and Clark, *Academic Charisma*, 74–80, 145–56.

64. *Atti della nazione germanica*, 1600, vol. 2, 171: "[8 January] Ingruentibus iam magnis frigoribus, coeloque iam nives, pruinas, glacies que demittente, Excellentissimus Aquapends absoluta prius absolutissima in publico auditorio tam humanorum quam ceterorum animalium, ut et volucrium seu pennatorium ossium ostensione, postea quam tria extarent corpora seu subiecta, duo virilia, muliebre unum, ad sectionem solemni pompa cum fidicinibus ab Anatomisticis conductis accessit, eamque aliquot dierum spacio, frequenti sempre auditorum corona admodum evidenter administravit, et tandem post luculentam oculi dissectionem sectioni finem imposuit."

65. The increasingly close relationship between the university and the state forms something of a leitmotif in the history of the University of Padua. See, most recently, Clark, *Academic Charisma*; see also Dooley, "Social Control"; Kagan, "Universities in Italy"; and Sandro De Bernardin, "La politica culturale."

66. Tomasini, *Gymnasium patavinum*, 79.

67. For example, Vesalius's connection to the humanist circle of the *Infiammati* in Padua is treated by Andrea Carlino in "Le fondements humanists."

68. Weinberg, *History of Literary Criticism*; and *Trattati di poetica e retorica*.

69. Howell, *Poetics, Rhetoric, and Logic*, 51.

70. While poetry affects the spectators by producing fear and pity (insofar as they see themselves in the characters of the drama) and then catharsis (the purging of that fear and pity), rhetorical works seek to persuade the audience. In his *Rhetoric*, Aristotle explains the complex reaction inherent in persuasion as an accumulation of rational beliefs about the thesis, emotional acceptance of the thesis, and an ethical consideration of the speaker. Persuasion is triggered "by a rational belief in the truth of the orator's thesis, by an emotional acceptance of the thesis as in some way pleasurable, and by an ethical acceptance of the orator's character as that of a man of good sense, good morals, and good will" (Aristotle, *Rhetoric*, II.55).

71. Giason de Nores, "Breve trattato dell'oratore," in *Trattati di poetica e retorica*, 106: "Nè dovemo dar audienza a coloro che affermano comunemente noi essere prodotti dalla natura et a lusingar apertamente, et ad entrar occultamente con umiltà nell'altrui grazie, et a spaventar gli adversarii, et a raccontar il successo del fatto, et a confirmar le nostre ragioni, et a rimprovar le contrarie, et ultimamente a conchiuder o con prieghi o con querelle, non considerando essi l'arte essere una certa osservazione et imitazione della natura, mirando noi infiniti far tutto ciò parte a caso e senza regola, parte sempre con ordine e con avvertimento, la qual cosa non dovemo per alcuna maniera pensar che possa essere eseguita senza l'aiuto dell'arte."

72. There was a long-standing tradition behind this connection, developed in the context of societies such as the *Infiammati*. See Lowry, "Proving Ground"; and Daniele, "Sperone Speroni."

73. Meduna, *Lo scolare*, 65v–66r: "perche all'huomo propriamente si convenga giovare all'huomo, da che ogni cosa da quel supremo Artefice, e Maestro per l'huomo fu creta, e l'huomo per porgere aiuto all'huomo è creato, ed egli in questo deve impiegar tutti i pensieri, ed tutti gli studi imitando la granmadre Natura, che à man piene i suoi benefici ed suoi favori sparge in commune."

74. Cremonini, *Il nascimento di venetia*, dedication: "Aristotele hà scritto nel la sua *Poetica*, che la Poesia è da Filosofo; nè Euripide, ne Sofocle non furono Filosofi; nè Homero è mancho di Filosofo, dell'autorità del quale si spesso si vale Aristotele. Hà il medesimo Aristotele poetato, e vive ancho qualche suo Poema. Fece lo stesso Platone, che compose Tragedia; et Empedocle e Lucretio poetarono la loro Filosofia, et affermano Platone, et Aristotele il poetare esser forza d'ingegno; la quale sotto nome di furore divino è proposta da Platone. Per il che non è se non da esser più riputato chi essendo buon filosofo è appresso buon Poeta."

75. Fabrici, *De formato foetu*, preface.

76. *Trattati di poetica e retorica*, 40–41.

77. See chapter 2. In practice, Casseri offered a lot of instruction on surgery. See Sterzi, "Guilio Casserio."

78. Portenari, *Della felicità di padova*, 96–97.

79. Schott, *Itinerario*, 42: "Et è cosa celebre l'Anfiteatro Anatomico drizzato in esse Scole ad uso de'Professori di Medicina: è lo Studio di Padova un famosissimo mercato delle Scientie, non altrimente, che si fosse anticamente l'Academia d'Athene." This passage is reprinted in *Padova: Diari e viaggi*, 32.

80. Evelyn, *Memoirs of John Evelyn*, 170.

81. *Atti della nazione germanica*, 1597, vol. 2, 110: "quamvis enim initio fervidissimi appareant, subito tamen denudato vix umbilico conspectisque abdominis saltem partibus, illico frigescere et subtrahere se incipiunt, paucissimis exceptis, quos pudor et amittendi apud Doctorem metus favoris invitos ad coronidem usque retinet."

82. *Statuta almae universitatis*, bk. 2: "De anothomia singulis annis fienda (xxvii): Ut autem res ordinate, & cum omnimoda utilitate procedat: rector cum sapiente, & consiliariis cum talis anotomia facienda est, eligant duos scholares idoneos, qui ad minus: in hoc studio per biennium in medicina studuerint, & si haberi possint qui viderint alias anothomias vocenturque; massarii anothomiae. Eorum sit officium de loco de instrumentis, & de omnibus necessariis providere, & taxare quantum quisqu; volens videre solvere debeatur, taxetur autem pro quantitate expensarum faciendarum. Ad eam autem videndam nullus scholaris nisi matribulatus, & qui medicinae ad minus per annum studuerit admitti possint."

83. *Atti della nazione germanica*, 1597–1598, vol. 2, 111.

84. Ibid., 108: "sed alios honestas, alios pudor refraenabit, spero."

85. From 1587 on, the records of the transalpine nation list the books that were donated to the library and, typically, the person donating them. The vernacular works included Italian language dictionaries, works on the language debate, Cicero's familiar letters, Boccaccio's *Decameron*, ancient and contemporary comedies, and travel narratives. For a list of books and donors, see *Atti della nazione germanica*, 1586 and following. On university libraries, see Rossetti, "Le biblioteche delle 'nationes,'"; and Grendler, *Universities*, 505–6.

86. Brugi, *Gli scolari dello studio*, 19, 86–91.

87. Aristotle, *Rhetoric*, II.12.165–66.

88. For a historiographical introduction to the history of education, see Black, "Italian Renaissance Education." For the classical study of Italian education, see Garin, *Il*

*pensiero pedagogico*; and for the most recent response to this tradition, one that explores the gap between theory and practice, see Grafton and Jardine, *From Humanism to the Humanities*.

89. Guazzo, *La civil conversazione*, vol. 2, 60 n. 114. Bartolomeo Meduna repeated this idea, explaining that a person was civil (or noble) if he exercised himself in virtuous and proper [*honeste*] actions. This was one role assigned to the teacher and the educational setting: to make virtuous and honest action habitual. See Meduna, *Lo scolare*, 66: "qualunque persona civile, e noble deve di continuo essercitarsi nelle virtuose, ed honeste attioni."

90. As humanists, the authors of these manuals emphasized a student's development by way of classical patterns of organization. Guazzo, for example, used being, nutrition, and discipline to structure his discussion (bk. III in Guazzo's text; vol. 1, 233 in Quondam's critical edition). The threefold structure recalls Plutarch's *De liberis educandis*, delineating *physis* (related to natural inclination), *logos* (interpreted variously as reason, doctrine, and discipline), and *ethos* (associated with instruction and intellectual exercise). For example, in *De educatione liberorum et eorum claris moreibus libri sex*, Maffeo Vegio explained: "Come si suole delle scienzie e de l'altre arti dire nel generale, così medesimamente è da dire della virtù, cioè che tre cose sono necessarie al compimento e alla perfezzione d'una operazione: la natura, la ragione e la consuetudine. Quel che chiamo ragione è la disciplina istessa, e la consuetudine è lo essercizio che si fa in quella cose; la disciplina è capo e principio del tutto, e l'uso si acquista mediante l'essercizio e l'operazione, e da tutte queste cose nasce la perfezzione" (vol. 1, 381–82). Other writers, such as Bartolomeo Meduna, emphasized the importance of exercise in learning discipline, effectively switching the second and third categories: "for nature, one considers that easily the concepts, when heard, are apprehended and with the impression, strongly retained; in exercise, with work and promptness and diligence, [the concepts] are used and adopted by the intellect; and in discipline . . . with habits that accompany wisdom . . . [they are] conserved and maintained." See Meduna, *Lo scolare*, 11: "tre parti sono molto necessarie allo studioso la natura, l'esercitio, e la disciplina, nella natura si considera, che facilmente i concetti uditi apprenda, e gli apresi fortemente ritenga, e saldamente; nell'essercitio, che con fatica, sollecitudine, e diligenza usi, ed adopri l'intelletto, nella disciplina, che lodevolmente vivendo i costumi con la scienza accompagni, la onde per conservare, e mantenere la natura, e per porre in uso l'essercitio, e la disciplina vi [b]isognamo l'educatione, e l'institutione."

91. Guazzo, "Conversatione insegna più che i libri," *La civil conversazione*, vol. 1, 11.

92. Ibid: "Et voglio dirvi di più, che sarebbe errore il credere, che la dottrina s'acquisti più nella solitudine fra i libri, che nella conversatione fra gli huomini dotti . . . che meglio s'apprende la dottrina per l'orrecchie, che per gli occhi, e che non accaderebbe consumarsi la vista . . . e ricever per l'orecchie quella viva voce, laquale con mirabil forza s'imprime nella mente . . . Io dopo vengo considerando, che l'animo del solitario diviene o languido, e pigro, non havendo chi lo stuzzichi col ricercar la sua dottrina, e col disputare."

93. For a summary example of a disputation, see Grendler, "Universities of the Renaissance," esp. 16, which shows a reprint of an announcement of a disputation on a

medical question by Mattheus de Montibus Blanchorum, a student, from 19 March 1508 in "Dispute e ripetuzioni di scolari per ottenere letture d'univesità 1487–1515," *Riformatori dello Studio*, filza 279r, Archivio di Stato di Bologna.

94. Paul Grendler notes that the skills needed to be successful included the ability to draw distinctions from logical principles, to state views forcefully, to isolate the errors in the opponents' statements, and to quote authoritative texts from memory. See Grendler, *Universities*, 153.

95. Meduna, *Lo scolare*, 96–97: "Bernard soggiunse. Non è per certo cosa, che più nella mente, imprima quel, che s'ha udito, e letto, che il communicare ed il disputare; onde per inanimare gli scolari alla sollecitudine degli studi con la lode, che si riporta ne gli spettacoli delle dispute, voglio succintamente favellare delle utilità di quelle. Chi dunque non vede, che per capire, e penetrar minutamente le sottigliezze, e l'argutie, non è che più vaglia della disputa, la qual fù sempre dai filosofi amata, i quali chiaramente conobbero, nulla esser più giovevole per acquistar la cognition della verità, che il continuo questionare, perche si come per la ginnastica, o per la lotta le forze del corpo diventano più ferme, e stabili, cosi nella battaglia delle lettere le virtu dell'animo si rendono più forti e vigorose . . . Poiche la disputa fa vivo l'ingegno, tenace la memoria, pronta la lingua, ardito l'animo, e chiara la verita."

96. della Casa, *Il Galateo*, 63.

97. Ibid., 58–59.

98. Ibid., 24.

99. Lorraine Daston treats this aspect in Francis Bacon's writings and in relation to the Royal Society in England. See Daston, "Baconian Facts."

100. Meduna, *Lo scolare*, 85–86: "Adunque il nostro scolare senza far molto alcuno con tutta la mente, e con tutto l'animo non fabricando castelli in aria, ne vagando co'l pensiero udirà il lettore attentamente pendendo tutta via dalla sua bocca, ed oltre l'ascoltarlo con maraviglia, e volontieri prestarà intera fede a cio che egli dice, e lo conservarà nell'arca della memoria, e se per sorte non havrà bene intesa la lettione, potrà con bella occasione ricercarlo, overo per non molestarlo à tutte l'ore essendo pupillo dimandarà l'intelligenza delle questioni, e l'importanza delle ragioni ai provetti: La soggiettione dello scolare si deve particolarmente ritrovare in tre cose . . . nell'attentione, nella docilità, e nella benivolenza; attento con l'essercitio, docile con l'ingegno, e benivolo con l'animo, attento all'udire, docile all'intendere, e benivolo al ritenere."

101. Ibid., 99–100: "All'ora lo scolare deve parlare, quando l'occasione l'invita anzi sprona, quando non gli è utile tacere, e quando a lui tocca, e per venire al come, voglio, che mostrandosi accorto, e prudente favelli saggiamente, veramente, e modestamente, e non pazzamente, simulatamente, ed arrogantemente, e se il ragionamento è grave, servi parimente gravita, e se è piacevolezza, consideri, che quello, che egli reputa esser male, e vituperoso ad operarlo, deve anco reputar sconcio, e vergognosa a dirlo."

102. Ibid: "Non meno voglio, che non interrompa gl'altrui parlare, e che non morda alcuno co'l motreggiare, e che non s'apponga ad ogni parola, e che non contrasti, e che non disputi con asprezza, e con mal garbo, e che non tenti con arroganza superar gli altri, ne dica il falso, perche meglio è dicendo il vero esser vinto, che dir la bugia vincere altrui, ne che sostini sposando, come si dice, la sua opinione; ma volontieri ceda, ben

c'havesse ragione, avvedendosi, che il compagno non volesse acquetarsi, accioche da i detti non venissero ai fatti e di amici si facessero nemici, percio mirando con cui prattica, o si sforzera di con farsi con quel tale, overo non terra la sua compagnia."

103. This demonstration was the one discussed at the beginning of this chapter.

104. *Atti della nazione germanica*, 1589, vol. 1, 269: "Nam in anatomes lectione publica, cum in dissectione musculorum linguae illarumque partium voci atque pronunciationi servientium versaretur, occasione hinc incepta sermonem satis tamen ridiculum de variarum Nationum pronunciatione habere coepit. Inter caeteras autem praemissas Nationes, ad laudatissimam quoque nostram devenit, dicens: Germanos durae atque tardae pronunciationis esse, siquidem os nimium, cum pronunciare vellent comprimerent, quod in causa esset, ut semper inepte literam f pro v pronunciarent, uti hisce annalibus verbis id demonstrare conatus fuit: qui ponum finum pipit, tiu fifit."

105. Ibid.: "Haec verba peroptime huic rei quadrare videns, non unica atque altera, sed crebrius et ad nauseam usque protulit, atque ex istorum verborum repetitione huiusmodi voluptatem persensit, ut risum simul cohibere non posset, quasi vero Aquapendenti alia verba defecissent, quibus modestius, si quod virum loquelae nostrae dicendum adsit, corrigere potuisset. Sed hinc facile coniiendum, id praemeditato studio factum fuisse, quo nos Germanos, ut insignes vini potatores, in tot Nationum praesentia ludibrio atque risui exponeret."

106. Ibid.

107. Ibid.

108. Lombardelli, *Gl'aphorismi scholastici*, 5: "Onde l'opposito [I]dee concludersi: perche lo Studio frena le passioni; abbassa l'orgoglio; ammollisce la fierezza; addolcisce i costumi; insegna la modestia; tempera la furia; modera i pensieri; esorta a viver contento del poco; nobilita la rozezza; placa gli sdegni; rende l'Huomo affabile; palesca, e difende la verità; compone le mentis; fa discernere i veri beni dagli apparrenti; in somma riduce la parte bestiale dell'Huomo ad un viver non pure humano ma quasi angelico."

109. Crispolti, *Idee dello scolare*, 60: "Oltra i detti significati, dimostra il libro aperto, che'l dottore [i]dee interpretare i libri e sforzarsi, quando sono ambigui, di ridurli ad ottimo senso, massimamente quelli de'gentili: e ancora correggerli con modestia, senza passione, e interesse."

110. Ibid., 26–27: "Lo scolare è quegli, che alle scolastiche discipline ha buona dispositione. Circa la qual dispositione due cose considerar si deono. L'una il dono della natura, il quale non è perfetto: L'altra il mezo per ridurlo à perfettione. Il dono della natura è detto docilità . . . laqual docilità si ricerca allo scolare: et in tre cose consiste, ingegno, giuditio, e memoria, le quali immediatamente dipendono dalle facoltà dell'anima, e principalmente dall'immaginatione, intelletto, e memoria: e queste facoltà hanno gran dipendenza da gli organi, e complessione, che assai deriva da i genitori, dalle nutrici, dall'aere, e dalle cause superiori. L'ingegno è una certa acutezza nell'apprendere."

111. Ibid., 28: "Veggiamo, che gli animali di fiera natura con la diligenza divengono mansueti, e tratabili; Non altrimente avviene de gl'ingegni rozi, e aspri, che con lo studio, e industria, s'inteneriscono, e si rendono idonee alle scienze."

112. See the recent work of Bloom, *Voice in Motion*; Horodowich, *Language and Statecraft*; and Mazzio, "Senses Divided."

113. "Sulle 'Cautele dei medici' di Alberto de'Zanchariis," in *Di alcuni trattatelli*, 13–14: "Il miglior partito in tal caso è di stare in silenzio e di fare l'orecchio del mercante, secondo il proverbio: infatti è meglio tacere anzi che errare, come dice Galeno nel proemio del Meracriseos, nel 2 cap . . . Rimane da discutere del compenso che deve ricevere il medico dal malato, poichè il medico deve sollecitare affinchè sia pagato secondo il suo diritto."

114. Gabriele Zerbi ends his chapter on assistants with an account of demographic differences: "Not every nation is fitted for this work. The English are men of immense pride, the Swiss intolerably suspicious, the Illyrians foul-mouthed, the Hungarians hostile to the Italians of all people. Ideal for this kind of work are the Bretons, Germans, some of the French, those Spaniards who are more similar to the Italians. Best of all the Italians are the Lombards." See Zerbi, *Gerontocomia*, 89.

115. Zerbi, *Gerontocomia*, 90.

116. *Atti della nazione germanica*, 1595, vol. 2, 58: "Confluxerat eo tota quasi civitas, et extremae etiam farinae homines tanquam ad forum cupedinis: subsellia occuparunt hebraei, sedentarii magistri, sartores, calceolarii, solearii, carnarii, salsamentarii et his inferiores baiuti et corbuli illi."

117. Ibid., 60–61. The full passage is cited, translated, and discussed in chapter 5, n. 10.

118. Ibid.

119. Ibid., 1596–1597, vol. 2, 88: "Ceterum anatomen integram brevi temporis intervallo tam praeclare confecit, ut non iniuria quispiam dicere potuisset, tantum illum hoc anno se ipsum superasse quantum ceteroquin alios antecedit anatomicos."

120. Ibid., 1597, vol. 2, 111: "A.d. 10 Xbris Celeberrimus Anatomicus noster, postquam Anatomistae duo procurassent corpora, virile et muliebre, suspensa foetuum, brutorum anatomia, humanam foeliciter et luculenter exorsus est."

CHAPTER 4: MEDICAL STUDENTS AND THEIR CORPSES

1. This chapter develops some of the concerns raised by Andrea Carlino in his chapter on "Knowledge and Ritual" in *Books of the Body*, 187–225. With reference to grave robbing, see Carlino, *Books of the Body*, 98 n. 98; and Park, "Criminal and the Saintly," esp. 18.

2. In his diary, Felix Platter mentions several trips to the cemetery to dig up freshly buried corpses. See Platter, *Beloved Son Felix*. For examples of the importance of grave robbing in later periods, see Richardson, *Death, Dissection, and the Destitute*; and Sappol, *Traffic of Dead Bodies*.

3. *Statuta almae universitatis*, bk. 3, ch. 28: "quod singulis annis fieret anatomia, quodq; pro ea perficienda rectoris urbis, & territoriiq; tenerentur cadaver cuiuscunq; delinquentis, de quo capitis supplicui summeretnr [sic], dare anatomiae deputatis, nisi cadaver esset alicuius civis veneti, aut patavini, quum iam multis dicta anatomia, nisi raro facta fuerit cadaverum defectu. Ideo utilitati non modo scholarium, sed etiam universo mortalium generi consulentes, non in aliquod quorum vili pendium, statuitur quod urbis prasides, ac omnium locorum Patavini districtus praetores teneantur dictis nostris anatomistis dare quodcunque cadaver cuiuscunque delinquentis capitis supplicio pun-

iti, nisi sit venetus, aut Padaunus, civis, veli ex comitatu ex aliqua familia alicuius aesti-mationis, & nisi consanguinei eius ex eadem familia contradicant. vel advena nobilis, vel alicuius existimationis, sub poena in dicto statuto contenta. Ad quam videndam omnes scholares nostri matriculati possint intrare, dummodo solverint, non obstante dicto statuto, quae solutio non possit excedere summam marcellorum trium argentorum, ex quibus fiant impensae necessariae, & praefertim exequamur." For references to Bolo-gna, Pisa, and Rome, see Carlino, *Books of the Body*, 77–91.

4. *Statuta almae universitatis*, bk. 3, ch. 28.

5. On the role of these authorities, see Carlino, *Books of the Body*, 92–98.

6. Park, *Secrets of Women*; Park, "Criminal and the Saintly"; and Sawday, *Body Em-blazoned*, 85–140. On burial, see Park, "Life of the Corpse." On burial practices in Re-naissance Italy, see Strocchia, *Death and Ritual*, esp. 5–54.

7. This was based on the reformed statutes of 1502, transcribed in de Sandre Gaspa-rini, "La confraternità," and cited by Park, *Secrets of Women*, 346 n. 9.

8. In his diary, Felix Platter describes several of these, suggesting a marked differ-ence in burial rites for northern and southern Europe. Platter, *Beloved Son Felix*.

9. In addition to de Sandre Gasparini, "La confraternità," see Prosperi, "Consolation or Condemnation."

10. Park, "Criminal and the Saintly," 12–14.

11. Ibid.

12. While the criminal proceedings and the correspondence between Paduan and Venetian authorities regarding Marco's case have not survived, the details of his ghastly crime are included in Niccolò Rossi, *Annali di Padova*, 193, MCP: "Principio quest'anno con un caso degno di compassione per lo crudeltà del delitto fatto per mano di Marco Fruitarolo giovine da 23 anni nella persona di Cecilia sua Moglie questo fu alli 13 Marzo mentre che habitava a nella Contra di Pontecorvo all'incontro delli SS. Fregozi. Et quan-tunque che questo fusse giovine e consapevole dell'impudicitia della giovine moglie bella; e che gli havesse molte volte dato comodità di ritrovarsi con qualche suo amante . . . Persuaso da diabolico pensiero lo sera circa alle tre hore di notte chiamato a se essa sua Moglie le comandò che andasse di sopra in una soffitto dove teniva diverse quantità di frutti . . . con la giovine non sospettando cosa alcuno del Marito vi ando come era solita altre volte par simili occasione di andarsi . . . Et egli poco dopo andò da lei e ritrovandola, con un manarino la cozzò senza strepito alcuno, e dopo fatto questo la spogliò nuda e la tagliò, e la squarto in otto pezzi.". Rossi's chronicle has yearly entries, which cover a wide range of topics: the opening of new churches and buildings in Padua, new installations of paintings and sculptures in those buildings, the fights between university students, the hiring and retiring of professors, the death and burial of eminent professors such as Fabrici, the arrival of important dignitaries from Venice, and reports about the papal in-terdict, as well as marvelous births, comets, dangerous fires, and crimes. The topographi-cal appearance of the manuscript suggests that Rossi wrote once or twice annually rather than daily or monthly, selecting (from memory or notes) the *novità* that he felt warranted recording. The range of topics implies more than a cursory knowledge of various com-munities in Padua, including the academic, artistic, monastic, and political ones. On homicide and the statutes of Padua, see *Statuti del comune di Padova*, 290–92.

13. The criminal's body was usually hanged before quartering. Exceptions were made for particularly brutal or brazen killers, who might be quartered before hanging. See Terpstra, "Theory into Practice," 130–31. According to Master Franz Schmidt, for public executions in Nuremberg from 1573 to 1617, it was unheard of for a woman, once convicted, to be hanged—until 1584 and the execution of repeat offender Maria Kürschnerin of Nuremberg, alias Silly Mary. Beheading was more common (and beheading with a sword was "a favor"). See *Hangman's Diary*, 125.

14. Rossi, *Annali di Padova*, 193–94, MCP: "li quarti davanti li portò nelle fosse di Pontecorvo giù dalle mure, gl'Interiori li portò dro S. Massimo al Porte vecchio nella Brenta le fosse le gittò la mattina nelli condotti del Palazzo."

15. Ibid: "Mandò la notte sequente alla casa del Fruttarolo il quale preso e messo alli tormenti confessò il delitto, il premio fu che fu condotto per la Giustitia alla casa della sua habitatione, et in quel loco per il Ministro della Giustitia li fu tagliato lo mano dritto e poi taglielo al collo fu ritornato alle Piazze dove sopra an eminente forca pagò il fio del suo scellerato misfatto con gra concorso di Populo tre giorni dopo l'amazzamento, e menato a coda di cavallo il corpo suo fu dato per anotomia alli Scolari."

16. Manutio, *Eleganze*, 13–14: "*Amazzare*: Oreste amazzò, uccise, privò di vita di sua propria mano la madre Clitemnestra. Orestes parete Clitemnestram sua manu defodit, confodit, perfodit, vita exuit privavait, morte affecit, punivit, multavit, ultus est, vindicavit, interfecit, interemit, occidit, cecidit, peremit, trucidavit, obtruncavit; parenti vitam eripuit, mortem obtulit, attulit, intulit, vim intulit, attulit, manus attulit, intulit." The transalpine students acquired this book for their library in 1591. See *Atti della nazione germanica*, 1591, vol. 1, 307.

17. Dante, *Inferno*, 28.103–5: "E un ch'avea l'una e l'altra man mozza, / levando i moncherin per l'aura fosca, / sì che 'l sangue facea la faccia sozza."

18. In addition to Park, "Criminal and the Saintly," see Jeffrey Masten, "Is the Fundament a Grave?"

19. Andrea Carlino traces the historical evidence for the general "disgust" for dissection in the anatomical literature of the early period, with anatomists reflecting on the superstitions about the dead body as a contaminating object. Carlino notes that today we would call this an anthropological concern for the body. See Carlino, *Books of the Body*, 118–19, 219.

20. Park, *Secrets of Women*.

21. See Foucault, *Discipline and Punish*; Sawday, *Body Emblazoned*, 54–84; and Sugg, *Murder after Death*, 12–35. For a historical account of execution, see *Art of Executing Well*.

22. Prosperi, "Consolation or Condemnation," 115.

23. Park, *Secrets of Women*, 213. On the broader development of this theme, see Ariès, *Hour of Our Death*, 40–45.

24. Prosperi, "Consolation or Condemnation," 115.

25. Edgerton, *Pictures and Punishment*. His book includes a number of photographs of a *tavoletta* belonging to the brothers of San Giovanni Decollato in Rome.

26. Andrea Carlino, "Dissection as Social Drama in Early Modern Europe," in *Spuren der Avantgarde: Theatrum anatomicum* (Berlin: De Gruyter, forthcoming).

27. Park, "Criminal and the Saintly," 23–29.

28. Giuseppe Sterzi records the first use of the word "anatomist" being applied to the student assistant in 1597, an oversight in what otherwise remains an exemplary piece of scholarship. See Sterzi, "Giulio Casserio."

29. Elsewhere, Fabrici clearly devoted himself to the labors of dissection, producing a series of colored illustrations of dissected animals. See *Il teatro dei corpi*.

30. *Atti della nazione germanica*, 1599–1600, vol. 2, 170: "28 Novembris habita est Universitas, in qua electi sunt Anatomistae, quos vulgo Massarios vocant, qui publice anatomiae administratori in secandis et separandis corporibus auxiliares manus offerre solent. Electus autem ex nostra Natione fuit Iacobus Gomans Belga. Ubi posteris hoc loci animadvertendum venit, ne hoc priveilegium aliquando amittant, neque Italis hac in parte manus dent, cum saepius Germanum Massarium, nescio qua moti iniquitate, admittere opposuissem, nostrum Massarium non admisissent."

31. Ibid., 1600–1601, vol. 2, 180: "11. d. Novembris Celeberrimus noster Anatomicus, Dominus Hieronymus Fabritius ab Aquapendente, absoluta prius in auditorio methodo anatomica generali, deinceps in Theatro anatomias foetuum, et quidem equini et agnelli aggressus est. 20. d. Novembris, convocata Universitate Magnifica, electi sunt Anatomistae, quos vulgo Massarios vocant, qui Excellentissimo Anatomico in dissecandis et procurandis cadaveribus auxiliares manus ferre solent. Proposui ex nostra Natione Doctissimum Dominum Hectorem Selanofium Hessum, qui cum duobus Italis huic muneri praefectus est."

32. Park, "Criminal and the Saintly," 13 n. 41. Adriano Prosperi notes the limited attendance at the funerals of criminals and believes that this lay behind the founding of confraternities, which sought to comfort the condemned. See Prosperi, "Consolation or Condemnation."

33. Cremonini, *Le orazioni*, 37–39: "Percorrendo il mondo con tale indagine rivolta fuori di sè, senza neppure compiere molta fatica riuscirà a scrutare a fondo come è fatta la realtà: in virtù di quale meravigliosa maestra cioè la natura—come il mitico cameleonte, per il quale non v'è nessun colore che esso non sappia assumere—incessantemente si vesta ed incessantemente si spogli di forme diverse; in virtù di quale capacità essa, stimolata da semi piccolis simi, sia in grado di produrre cose grandissime; in che cosa la quercia sia superiore all'oro, in che cosa il leone sia superiore alle querce, e l'uomo al leone; in grazie di quale indissolubile e venerando legame di amore tutti gli esseri naturali sono uniti . . . dalle cose infime si possa salire alle eccelse attraverso armoniche graduazioni."

34. Ibid., 39: "contemplerà lui che non è pervenuto all'esistenza da alcun inizio temporale, che non è stato bambino né giovane e non sarà vecchio, ma susiste perpetuamente nella sua identità con sé, nella sua compiutezza ed ineffabilità, vivendo una vita felicissima consistente nella contemplazione intellettiva di se stesso, quale ci è possibile vagamente immaginare ma in nessun modo comprendere."

35. See also Carlino, *Books of the Body*, 78 n. 27.

36. *Atti della nazione germanica*, 1597, vol. 2, 110: "[3 December] In ea ipsa universitate, qua de referre coeperam, deliberatum quoque fuit numquid pro dissectis exequiae rursus solemnas uti antiquitus moris fuit instituendae, adque sumtus faciendos contibutio a spectatoribus exigenda? Propositum istum omnium calculis adprobatum fuit, permoventibus nos iniquioris vulgi rumoribus, dum passim proclamaremur, complura

cadavera raperemus inque Theatrum profananda comportaremus, crudeliter postmodum dilaniata non sepeliremus, sed vel in profluentem demergeremus, vel quod abomin-
dandum, etiam canibus non raro devoranda committeremus; et nescio quae non vana-
rum querelarum et lamentationum vanissimarum aliae fuerint."

37. Ibid.

38. Ibid: "Ergo latratuum istorum compescendorum caussa, tum etiam ut maiorem
cadaverum copiam nancisci possemus, sententiam de sepeliendis honeste imposterum
dissectis publice promulgare, et spectatores ad eleemosynas exhortari constituimus,
non tamen nisi consulto prius Anatomico, qui simul ac nostram persentisceret voluntatem, sibi istud negocium relinqueremus, nec suae auctoritati derogaremus satis morose
monuit; sique conatus istos qualiter cunque pios omnino praevertit. Quid ageremus? ne
enim ab incaepto desistat, blandiendum crebro ei et assentiendum minime vero reluc-
tandum esse, successores etiam experturos scio."

39. Tosoni, *Della anatomia degli antichi*, 87. See also Jacobo Facciolati, *Fasti Gymnasii
Patavini* (Patavii: Seminarii, apud Joannem Manfrè, 1757), vol. 2, 208–9 (cited by Park,
"Criminal and the Saintly," 18, n. 62): "Anatomicum studium in dies magis cum vigeret
non publicae modo, sed privatae quoque exercitationes, passim habebantur; quibus si
forte cadavera non suppeterent, ne sepultis quidem iuventus parcebat. Quapropter Sena-
tus consultum VI. id. febr. factum est gravissimarum poenarum sanctione adversus
illos, qui per huiusmodi causas sepulchra violarent."

40. See chapter 2, n. 50 and n. 51.

41. See chapter 2, n. 53.

42. *Atti della nazione germanica*, 1595, vol. 2, 59: "Videbatur iam vulnus solidum
quod recruduit a meridie: etenim parvi praetoritium hoc aestimans mandatum, quodam Mediolanensium in Universitate Legistarum Consiliarius, haud dissimili fastu
Massarios collegii nostri socios nonnullos, atque ex licita consuetudine sclopetis provi-
sos adgreditur, cui indignis incultisque verbis illos et Nationem nostram vellicanti, ex
nostris quidam bombardam monstrans, illine facessat rogat atque a maledictis linguam
temperet lubricam multiloquam."

43. Ibid., 58–59: "Quod postero qui fuit XII Kal. Febr. exigentibus Massariis, a iam
finita Excell. Panceroli I. C. lectione Foro-iuliacenses quidam omnium scholarium pet-
ulantissimi irruunt, qui quod pro libera anatomia solvant Germanorum solummodo
inventum identidem personantes, vi in theatrum irrumpere conantur: quibus vero
Massarii nostri cum impari fortitudine resistant, quodam audaciore ceteris et insolen-
tiore, pugno Domino Fabricio minitante, socii petulantior tantum impetum faciunt, ut
effractis quasi foribus voti fiunt compotes. Hanc petulantiam ac execrandum fastum
Anatomistae, qui indefessa diligentia ut rite et cum laude publicae sectionis administrationes in novo illo theatro procederent laborabant, aegerrime tulere."

44. Sterzi, "Guilio Casserio."

45. See chapter 3. The reference to tranquility comes from *Atti della nazione germanica*, 1597, vol. 2, 111: "A.d. 12 Xbris ad exhilarandos anatomiae spectatores recreandosque ex tristi aspectu animos, ex vetusta consuetudine (quae tamen superioribus
aliquot annis proximis interrupta) fidicines ab Anatomistis conducti et in Theatrum reducti fuere, procurante hanc rem sedulo D. Placotomo; aderantque musici isti etiam se-

quentibus diebus quamplurimis, sumtus certe qui illis irrogantur minime poenitendi, si quidem dum ipsis attendunt et auscultant spectatores, ab omni tumultu et calcitratione supersedere solent, cuius tranquillitatis gratia Theatrum anatomicum aliquot annos diutius inconcussum durare poterit." The passage is also cited by Gamba, "Il primo teatro anatomico."

46. On the relationship between students and clergy in Padua, see Brugi, *Gli scolari dello studio.*

47. *Atti della nazione germanica,* 1587, vol. 1, 246: "Hi igitur unanimi omnium consensu recepti et approbati fuere, hac conditione ut tria corpora seu cadavera illi procurarent, et pro quolibet a quovis anatomiam inspecturo 12 sol. exigerent. Si autem plura tribus invenirent, hoc honorem potius quam utilitatem ipsorum cedere, neque his 36 sol. quicquam plus exigere eos a scholaribus debere. Electi et confirmati iam erant Massarii, res ad Anatomicum defertur, qui nimium hoc esse praetium praetendens, alios qui pro 8 solidis pro quolibet cadavere solvendis hoc munus susciperent se inventurum promisit, atque ita post aliud quatriduum quatuor adolescentulos scholam Iesuitarum frequentantes, non modo huius exercitii omnino imperitos, sed et omnis medicinae penitus ignaros Universitati proposuit. Alia itaque universitas habetur, et ex malis quatuor, duos meliores eliguntur (quidem id ex actis Universitatis super hac re in aerario nostro repositis latius videre licet), hoc pacto ut 4 cadavera inveniant et mercedis loco in universum 32 solidos a quovis scholare, idest octo solidos pro cadavere accipiant."

48. Ibid.

49. Ibid.

50. Ferrari, "Public Anatomy Lessons."

51. *Atti della nazione germanica,* 1605, vol. 2, 228.

52. *History from Crime,* viii.

53. Ibid., x.

CHAPTER 5: PRIVATE ANATOMIES AND THE DELIGHTS
OF TECHNICAL EXPERTISE

1. On Casseri's relationship with Fabrici, see Sterzi, "Giulio Casserio."

2. *Atti della nazione germanica,* 1597–1598, vol. 2, 114: "Circa initium Februarii, postquam Eccellentissimus Aquapendens anatomiam maxima ex parte absolverat (de universali enim uti vocat ipse iam loquor, non de particulari et exactiori, quam nostro hoc anno de apprehensionis scilicet progressionis respirationisque organis ad 12m usque Martii deduxit), quidam nostri ordinis et Nationis homines haud postremi apud Excellentissimum Iulium Casserrum Placentinum obtinuere, ut is etiam (sicuti superioribus annis compluribus sponte fecerat) anatomiam in Germanorum otissimum gratiam habere denuo quam exactissimam promitteret." This document is transcribed and discussed briefly in Sterzi, "Giulio Casserio," 78 [pt. 2].

3. Ibid.: "Id quod revera praestitit, non tantum seorsim in simia a nobis oblata, qua plurimum excitatus fuit, sed et in vivorum canum aliquot apertione, tum vero in novem humanorum cadaverum consectione, quam quidem luculentissime maximo cum applusu et ingenti aliorum quoque Nationum concursu ultra 5 septimanas administrasse

ipsum nemini obscurum esse debet, praesertim quomodo integram totius animalis fabricam in tribus distinctis subiectis totidem horis (at prius diebus plus tribus non sua solum sed Germanorum aliquot opera rite praeparatis) maximo cum honore monstraverit."

4. Aspects of technical skill in the history of anatomy will be treated in Domenico Bertoloni-Meli, *Mechanism, Experiment, Disease: Anatomy in Malpighi and His Time* (Baltimore: Johns Hopkins University Press, forthcoming). In studies of the Scientific Revolution, technical skill has been addressed in artisanal culture and as a component of vernacular epistemologies. See Long, *Openness, Secrecy, Authorship*; Smith, *Body of the Artisan*; Harkness, *Jewel House*; and Dear, "Towards a Genealogy."

5. A complete account of the role of instruments in the history of anatomy is beyond the scope of this book, though it would illuminate some of the complexities of William Harvey's relationship to his Paduan medical education. Fabrici's interest in surgical instruments is developed in his two publications on surgery: *Pentateuchos Cheirurgicum* and *Opera chirurgica*. See especially Ongaro and Rippa Bonati, "L'Ortopedia."

6. Bertoloni-Meli, *Mechanism, Experiment, Disease* (forthcoming).

7. Lorraine Daston, "Attention."

8. In 1588–1589, the students said Paolo Galeotto gave "a thorough and complete anatomy, in which not only did he demonstrate most clearly and with amazing and beautiful ease and method the way to dissect bodies, the structure of all the parts and their actions and functions, but he also showed us the ways . . . through the whole body of the veins, arteries and nerves, to the great delight of us all." See chapter 2, n. 131.

9. For the full record, see n. 10.

10. *Atti della nazione germanica*, 1595, vol. 2, 60–61: "Neque eorum standum esse sententiae unquam et consulo et suadeo, qui liberam anatomiam quasi rei literariae pestem in hoc statu academico, atque fenestram seditionis ac caedis latissimam contra morem antiquum et ad nostram hanc usqu aetatem servatum, introducere conantur. Etenim quid nobilissimos Gymnasii celeberrimi studiosos, privilegiis singularibus praeceteris et immunitatibus gaudentes, commoveret vehementius aut ad arma excitaret celerius, quam invido conspicere oculo aut pati oscitantium etiam mechanicorum et proletariorum hominum extremae sortis catervam, subsellia immo universum theatrum occupare, fructum illum anatomiae iucundissimum conculcare quasi, minuere, impedire? Valeat, si sapiamus, valeat detestandus desidiorum fomes, et ansa lubrica, cum melius sit praesentibus exigua uti, quam frui incerta sperandis spe."

11. Ibid.

12. They continued to be called *anatomistae* from then on. See Sterzi, "Giulio Casserio," 235–37 [pt. 1]. Also see *Atti della nazione germanica*, 1597, vol. 2,110: "Nec solum Anatomistam, seu Massarium vulgo, ex Senioribus nostris quotannis habere cura sit, uti quem cum socio in procuranda cadaverum seu subiectorum copia, intromittendisque in Theatrum personis certis occupari novimus; quin immo si fieri potest, etiam collaborator seu Praeparator ex nobis quaerendus, quo celerius Senex progredi possit, nec habeat quod tarditatis excusandae loco umquam praetendere valeat."

13. *Epistolario della nazione artisti*, anonymous, letter to Johannes Richter Oppaviensis, June 1597, 141–42, AAUP: "Is enim Scholar Excellentium Virorum frequentando,

praxim medicam sectando, quaeque vel intellectu vel visu dignior occurrerunt diligenter persequendo, de gravioribus artis medicae controversii saepe cum aliis dissertando, demum anatomicis et chirurgias administrationibus operam indefessam . . . tantum ingenii atque eruditionis famam acquisivit ut omnium oculos animosque mi se converteret et singulare nationis Germanicae ornamentum a plurimis haberetur." Jerome Bylebyl noted that medical students, especially foreign students, sought practical training in clinical diagnosis as well as in surgery. See Bylebyl, "School of Padua." While Bylebyl's ideas are situated throughout the sixteenth century, my argument suggests that they emerged as a result of institutional and professional changes in the last two decades of the sixteenth century.

14. *Epistolario della nazione artisti*, 141, AAUP.

15. Ibid., 156–57: "exercitionibus item varias Chirurgicas et anatomicas."

16. *Atti della nazione germanica*, 1597, vol. 2, 94.

17. Ibid., 1596, vol. 2, 80: "sicque eorum ad quos pertinebat incuria vel etiam fortunae iniuria, maxime necessaria hac speculatione brumali privati fuimus, siquidem pollicitationum aurearum minime memor Anatomicus lectionibus de Tumoribus praeternaturalibus ac de Articulationibus residuum temporis tribuit . . . non tamen defuit studium Exc. Iulii Placentini, qui a nostris requisitus et exoratus, aliquot sectis cadaveribus, publicum dispendium privata diligentia recompensavit."

18. della Croce, *Chirurgiae universalis opus absolutum*, letter: "Cum igitur instrumentorum, quibus utiliter illi referti sunt, typos essem nactus, nihil habui antiquius, quam ut praelo subiecti quam accuratissime recuderentur."

19. *Atti della nazione germanica*, 1599–1600, vol. 2, 170: "ultimo novembris ego una cum Procuratoribus Domino Raab et Domino Dietero totius Nationis nomine Excellentissimum Aquapendentem accessimus, quid nobis hoc anno de futura anatomes administratione expectandum sit; rogavimus etiam ut in gratiam Germanorum, quorum satis copiosus confluxerat numerus, anatomicas administrationes mature et singulare quippiam, visuque dignum in medium proferre velit: id quod se facturum pollicitus ventris inferioris ostensionem, item chirurgicas administrationes, chirurgicorum instrumentorum exhibitionem absolutissimam certo esse visuros."

20. The first of Fabrici's publications on this topic, *Pentateuchos Cheirurgicum*, appeared in 1604; the second, his *Opera chirurgica*, appeared posthumously in 1647. The initial book was a catalog of conditions and treatments. In it, Fabrici combined a humanist study of classical works on surgery and an appraisal of the contemporary state of the field in order to situate the surgeon as a learned practitioner, one at the forefront of medical developments. His second book focused on questions of practice and "various modes" of operations, both ancient and "modern," including new ones invented and adapted by practitioners. The two works were republished together, in the vernacular, as *L'opere cirugiche di Girolamo Fabritio d'Aquapendente* in 1671.

21. *Atti della nazione germanica*, 1600–1601, vol. 2, 180: "postquam Anatomistae corpus procurassent, remota foetuum, brutorum anatomia, humanam exorsus est. Nec hac sola contentus, quamvis ex illa sat defatigatum se sentiret, resum[p]tis saltem paululum viribus, ad demonstrationes operationum chirurgicarum vir hic solertissimus deproperat; ac talem fidem ac solertiam in illis exhibuit, qualem repromereri non facile

possumus." In the same year, the university's administration tried to negotiate a private anatomy, but the record for this is limited to a marginal notation (ibid., 181).

22. Ibid., 1602, vol. 2, 193: "Adveniente tandem Studii publici tempore, quo virere vicissim atque reviviscere amoenissimi illi studiorum publicorum horti consuevere, cum aliquibus ex Natione nostra Excellentissimum Aquapendentem adii, ut quid spei de studio anatomico esset intelligerem: qui, non modo chirurgicas operationes (quas illico etiam exorsus est) pollicebatur, sed insuper anatomiam exactissimam promittebat, quam primum pristinae valetudini restitueretur; in lecto enim tum temporis ob morbum decumbebat."

23. Ibid., 1603, vol. 2, 195: "17 Martii Venetiis non multo ante reversus Aquapendens ad demonstrationes vicissim rediit suas, quas in theatro bene coeperat, nemine ad id hortante, nemine impellente, sed unico amore gratificandi studiosis, quo penitus hoc anno exarserat, adductus. Cupiebat quidem is chirurgicarum operationum ostensionem pertexere, quam utiliorem longe esse anatomiis praetendebat, sed studiosorum clamoribus dehortatus, qui anatomiam exposcebat, huic se vicissim accinxit, quam et diligenter persectus est, fideliterque absolvit; ut non nisi summe commendare Viri Excellentissimi industriam hoc anno adhibitam valeamus."

24. Benetti, "La libreria."

25. Sterzi, "Guilio Casserio," pt. 2, 82–83: "del presente anno, nel quale lo Ecc. Acquapendente tutti li giorni così solo ha tagliato la Anatomia, ma fatto tutte le operationi chirurgiche, mostrando come si debbono medicar le ferite, che forse in ciascun caso si debbino usar, comparando li antichi con li moderni, et fatto insomma tutte quelle cose, che a buon chirurgo si convengono, con grandissimo concorso et profitto di Scolari, et universal sodisfattione, cosa, che già molti anni et forse mai più è stata fatta in questo Studio. Onde si può con verità dire, che ha superato se stesso. Per far tutte queste operationi, hanno bisognato molti corpi, delli quali alcune volte è stato mancamento, et se non fosse stata usata da noi, et estraordinaria diligenza in favorir così buona et utile opera . . . Perche, essendo alcuni pochi scolari sollevati da un certo Dottor Piacentino, che non ha alcun carico dello Studio, ad distanza del quale rubavano dei corpi, apportavano molto disturbo et impedimento, si come ad instigatione di questo, crediamo, che sia bene, che non sia altrimenti confirmata, perchè di avantaggio è proveduto, che finita l'Anatomia pubblica si possin far le particolari con licenza, et sopra intendenza dei Rettori."

26. Ferrari, "Public Anatomy Lessons."

27. Renaissance surgery in Italy awaits a full study. Nancy Siraisi discusses the combination of intellectual and practical learning in the university, noting that practical skills could be acquired in formal and informal settings: in the university, in classrooms and private study, or in shared experiences. She also notes that between the fourteenth and sixteenth centuries, surgery flourished in the context of the guild, and a distinct hierarchy between surgeons and learned (university-trained) practitioners appeared. This professional separation was less apparent in Italian cities, in part because learned surgery had a place in the medical curriculum at Italian universities. See Siraisi, *Medieval and Early Renaissance Medicine*, 49–50, 174–79. On the late medieval tradition of rational surgery, see McVaugh, *Rational Surgery*. On Venetian physicians, surgeons, and medical

colleges, see Palmer, *Studio of Venice*; Palmer, "Physicians and Surgeons"; Bonuzzi, "Medicina e sanità"; and Bernardi, *Prospetto storico-critico*.

28. See Park, *Doctors and Medicine*, 8; Palmer, "Physicians and Surgeons"; and Siraisi, *Medieval and Early Renaissance Medicine*, 153–86. For historiographical perspective, Carlo M. Cipolla argued for the drastic distinction between the two. See Cipolla, *Public Health*, esp. 75–76. For a critique of Cipolla, see *History of Medical Education*, esp. 100. O'Malley believes the overstatement of this gap occurred on the basis of conflicts in London and Paris.

29. Richard Palmer explains that for a surgical degree, the statutes of the University of Pisa in 1478 required two years of university education in surgery plus one year of practical experience, indicating that academic training was available for Tuscan surgeons at least by that date, and precisely the same requirements were laid down in the statutes of the College of Arts and Medicine in Padua, as revised in 1607. See Palmer, "Physicians and Surgeons," 453–54.

30. On the emergence of a humanist tradition in surgery, see especially Nutton, "Humanist Surgery."

31. *Atti della nazione germanica*, 1597–1598, vol. 2, 113: "9 Ianuarii Doctor Tiberius Phialetus Bononiensis Chirurgus et Anatomicus, postquam tres tantum decubuisset dies, ex repentina paralysi diem suum obiit."

32. Ibid., 1599, vol. 2, 140: "Cuidam Iohanni Bleuvfuss Hasso Chirurgo paralytico ex nationis aerario elemosynarum loco decreti sunt coronati duo."

33. Ibid., 1606–1607, vol. 2, 262: "20 Nov Excellentissimus Adrianus Spigelius Bruxellensis, Med. D. ac Illustrissimae Nationis Germanicae medicus ordinarius, suum cui titulum fecit Isagoge in rem herbariam librum Illustrissimae Nationis Germanicae Iuris Medicinaeque studiosis dicatum, mihi, praemissa primum praestitorum sibi ab incluta Natione officiorum enumeratione ac gratiarum actione, tradidit."

34. Ibid., 1606–1607, vol. 2, 274. His last name has been lost, due to a lacuna in the text.

35. *Atti della nazione germanica*, 1610, vol. 2, 324. Della Croce, the son of a surgeon and hailing from Dorsoduro parish, was licensed by the college of surgery in Venice and accepted as a member in 1532. He spent time as a surgeon for the naval fleet and subsequently became prior of the medical college, where he entered a culture of learned, sophisticated, and urbane practitioners. A sign of that sophistication was his connection to Francesco Sansovino (1521–1582), with whom he worked on an Italian translation of Giovanni di Vico's (ca. 1450–ca. 1525) *Practical Surgery* (1560). Della Croce published two volumes on surgery, first in Latin (1573) and then in Italian (1583). He identified himself as a "medico venetiano" and offered "the theory and the practical aspects necessary in surgery," emphasizing his debt (as a humanist surgeon) to Hippocrates, Galen, Celsus, and Avicenna. In his essay "Humanist Surgery," Vivian Nutton has shown that classical texts on surgery were slowly made available in new editions, mostly during the sixteenth century. On Francesco Sansovino and the vernacular print culture of Venice during this period, see Bareggi, *Il mestiere di scrivere*.

36. *Atti della nazione germanica*, 1595, vol. 2, 60: "Itaque Excellentissimus Iulius Placentinus ad hoc munus rogatus, etsi initio praxi et privatis aliis se excusavit, postmodo

tamen foetu ipsi a nobis oblato, in nostratium gratiam rem aggreditur, atque sectionem cum circa vasa umbilicalia, tum etiam quo ad foetus generationem attinet felicissimo successu auspicatur. Quo absoluto, unanimi omnium adsensu, duo ipsi cadavera, virile ac femineum procuratum fuit, in quibus non solum quoque perfectam et absolutam anatomen administravit, sed etiam varias operationes chirurgicas summa laude et studio accuratissime demonstrabat."

37. The document is transcribed by Sterzi, "Giulio Casserio," 81 [pt. 2]: "Essendo adunque l'Anatomia una delle più utili, et necessarie letture di Medicina, che siano nello Studio, et pertanto essendo bisogno di vederla, et rivederla assiduamente, et non potendo particolaremente le Nationi oltramontane con molti altri scolari ancora fermarsi tanto nello Studio, che facciano in essa l'esperimento che si richiede, attendendovi solamente quel tempo, che l'Ecc. Sig. Acquapendente spende in detta Lettura, la quale è da lui trattata con ogni diligenza, esquisitezza, et dottrina; et non potendo il medesimo sig. per le gravi occupationi che ha, esser d'ogni tempo in ciò impiegato; Pertanto la nostra Università si ha preso parte di supplicare, come fa, humilmente Vostra Serenità degnarsi di concedere, che fornito il tempo delle vacanze, deputato dalli Statuti nostri alla publica Anatomia, li qualunque Dottore, per venir esercitando quello, che habbino imparato nella pubblica del sig. Acquapendente, et in più breve tempo imparar de quest'arte quello, ch'è necessario loro, per la medicina."

38. Ibid.

39. Ibid., 84–85.

40. *Atti della nazione germanica*, 1604, vol. 2, 211: "Magnifica itaque Universitas illud officium Excellentissimo Iulio Placentino demandavit, qui id libenter etiam in se suscepit, et octiduo post in aedibus suis anatomiam perfectam ac sufficientem nobis exhibuit." In March 1604, Fabrici complained and said he would propose a tract on the liver and would trace and read for the anatomy, which would be general and have "solemn additions." He wished to "adorn and honor" the anatomy, which would not be "scorned" (ibid., 212).

41. Ibid., 1605, vol. 2, 227: "Ianuarii Universitatem coegit V. Rector, in qua; 1. Conclusum, Campanae operam, quam superiori hyeme in anatomia praestiterat, scribendam in Actis: ne vero ipsi obessent literae quas Rector praeteriti anni relinquerat et in Acta inseri fecerat, in quibus testabantur consensum Universitatis soli Placentino concessum fuisse licentiam privatae Anatomiae; statuerunt illas annullandas partim quod consensum ille factus esset absentibus quibusdam Consiliariis, partim quod literae Actis inscriptae essent insciis omnibus."

42. The transalpine students waged arguments concerning an anatomical counselor, a student elected by his nation. In January 1613, the students urged the *riformatori* "to satisfy our petition that our Anatomical Counselor should not only be admitted but also confirmed in this and all following years." Sometime later, Casseri "exhibited the dissection of a human corpse in a private establishment, and afterwards Aquapendente learned of this, bearing it most grievously, he rendered difficult the whole business concerning our Anatomical Counselor, so that afterwards on this subject we could no longer sway the Illustrious Rectors, as naturally in this matter they were entirely dependent on the nod of the old man." The students asked Giambattista Da Monte to intercede on their

behalf and, when he did, he brought "all kinds of jewels" to Fabrici in order "to placate the old man and indeed urge him to move our business forward." The transalpine students recommend that their successors pursue this unfinished business and "obtain the right of selecting the Anatomical Counselor from the Nation." It is never made explicit what the anatomical counselor did, except that he was supposed to attend the public anatomy demonstration and stand near the pit of the theater with other prestigious members of the audience. Perhaps he was there to take notes. The position was so important that it was deemed necessary to have an alternate. On 7 January 1616–1617, as the transalpine nation was agreeing on the election of *massarii*, the records of the nation indicate that "D. Constantinus Weckerus had been confirmed as the anatomical counselor" but that subsequently he had been called home. The record concludes: "our feelings were torn in different directions, for which it was debated whether one or two [counselors] ought to be proposed, lest the nation afterwards fall into such a labyrinth of controversies, which was approved by the whole assembly with unanimous consensus. And in this way, D. Jacobus Achenius and D. Henricus Botterus were proposed, from whom, the votes collected, D. Achenius was elected and confirmed, who administered this office with his excellent skill." Important though he was, in 1617 the captain of the university threw the anatomical counselor out of the theater because, "not knowing what he was doing, [he] sat in the chair of the Captain."

43. *Atti della nazione germanica*, 1613, vol. 2, 363–65.

44. Ibid., 364: "25 Augusti, inito prius cum Dominis Procuratoribus consilio, ipsum accessi, at cum primum Nationem pristino eiusdem favori et patrocinio commendassem, de his duobus cum ipso coepi conferre, num videlicet a Serenissimo Dominio Veneto privatas habendi anatomias et promovendi sine iuramento licentia impetrari posse videretur, praesertim cum occasio urgendi hoc negotium, ex interdicto ne Comites Palatini suo in privatis aedibus promovendi iure in posterum uterentur, satis ampla nobis oblata esse appareret."

45. Cunsolo, "Giulio Casserio," esp. 391–93.

46. *Atti della nazione germanica*, 1613–1614, vol. 2, 369: "Quo quidem tempore postquam in vivi canis sectione nervos recurrentes ostendisset, ad spectatores se convertens gratias egit iisdem, quod hactenus tanto studio se se docentem audire fuissent dignati, atque beneficii quasi petiit loco, ut dissectorum corporum exuvias ad templum Servitarum eorumdem sepulturae destinatum comitari vellemus."

47. Ibid.

48. Giulio Casseri, *De vocis auditusque organis*, 5: "Sed antequam, huic manum admoveam tabulae, praemonendus videtur lector, definitivam hic, observari rerum tractandarum seriem; quae explicato nomine, rem ipsam per structuram partis, eiusque a finali causa de promptas utilitates praedam perpetuo methodum in eorum expositione; quae circa humani corporis partes versantur. Earum altera historica plane Activa dici potest, et operatrix, quae fabricam accuratissime pandit, et exquisitam adeo parit, vel minutissimarum particularum notitiam: ut ex hac singulae possint artificiosa sectione, non lacerae, et illaesae ad vivum separari. Ista vere Anatomica Methodus dici debet, ut perpulchre Galenus testatum reliquit primo de loc. Aff. 1 cuius nominis gratum scio, vel gravissimis viris fuisse, et nunc esse, abusum. Altera (quae prioris, apud illos insignitur

titulo) contemplativa omnino, et intellectiva, solo mentis acumine, neglecta manuum opera, ab iis quae insunt. Temperamentis, consequentibus, accidentibus, partium usus rimatur, et utilitates expendit; atque huius meminit Galen in de usu partium cap 9 quae, cum in hoc satis elucescant opere, longo sermone lectoris tempora morari inutile duco; meque ideo ad Laryngis historiam accingo."

49. Ibid., dedication, and letter to the reader.

50. Ibid.: "ut quae longae experientia, et contemplatione observassem."

51. Ibid., 4–5: "et cum primis Philosophis, qui ex rerum mirabilium admiratione philosophari coeperunt; ego etiam diu multumque philosophando, perfectam laryngis mihi comparare notitam, totis semper animi viribus, nervisque contendi. Quam vero diuturna contemplatione sum affectus, hanc pro totius humani generis commodo, aperire iam gestio. interim unice petens, ut si forte non omnibus numeris absoluta accipiatur tractatio, aequi hoc bonique consulatur; et me calcar aliis, ad diligentius contemplandum peritiusque rimandum addidisse statuatur."

52. Ibid., ch. 2.

53. Cunsolo, "Giulio Casserio," 393–401.

54. Casseri, *De vocis auditusque organis*: "Sed antequam, huic manum admoveam tabulae, praemonendus videtur lector, definitivam hic, observari rerum tractandarum seriem; quae explicato nomine, rem ipsam per structuram partis, eiusque a finali causa de promptas utilitates praedam perpetuo methodum in eorum expositione; quae circa humani corporis partes versantur."

55. *Atti della nazione germanica*, 1613–1614, vol. 2, 369: "3 Ianuarii anatomen suam Excell. Placentinus exorsus est, publicam quidem illam, at non in theatro publico, ut supra dictam, sed in aula Ill. Praefecti, in qua maximis suis sumptibus theatrum in loco satis commodo extruendum curaverat, illamque ipsam Excell. Quorumquam vivorum condecoratam praesentia, magnaque studiosorum caterva frequentatum, tum ob honori suo, tum vero ut utilitati inserviret publicae, concurrente insuper singulari Massariorum in curandis cadaveribus industria, usque ad 14 mensis Martii diem absolutissime exhibuit. Quo quidem tempore postquam in vivi canis sectione nervos recurrentes ostendisset, ad spectatores se convertens gratias egit iisdem, quod hactenus tanto studio se se docentem audire fuissent dignati, atque beneficii quasi petiit loco, ut dissectorum corporum exuvias ad templum Servitarum eorumdem sepulturae destinatum comitari vellemus."

56. Ibid. Additional information supposedly appears in Franzina Bartolomeo, *Collaudatio mortuorum quorum anatomen publice professus est Patavii perillustris et excellentissimus Iulius Casserius Placentinus, Doctor Medicinae, Eques Divi Marci et Chirurgiae in celeberrimo Gymnasio Patavino Professor Publicus, dum funus honorificentissime fieret* (Padua: Laurentii Pasquati, 1614).

57. *Atti della nazione germanica*, 1613–1614, vol. 2, 369.

58. Casseri, *De vocis auditusque organis*, ch. 8. This translation is from Casseri, *Larynx, Organ of Voice*, 29–33.

59. Bertoloni-Meli, *Mechanism, Experiment, Disease* (forthcoming): "In his *Discours* on the anatomy of the brain, Steno argued that it is generally impossible to understand how a machine is assembled by observing its motions because those same motions

could be performed in different ways; rather, it is necessary to take it apart and examine all its minute components—*ressorts* is his term."

60. This example may also suggest an early figuring of what Bertoloni-Meli calls localization, the desire to localize function by studying the structures and microstructures that produce it.

61. This testimony is transcribed by Sterzi, "Guilio Casserio," 86–87: "[23 April 1604]: Hora essendo da questi inteso quanta utilità habbia apportato l'havere letta ed ocularmente mostrata questa Anatomia, così per la sua diligenza et cura usata, come per la comodità che dava ad ogn'uno in farli vedere particolarmente tutte quelle parti che in così breve spatio di giorni si potessero mostrare, e vedere, non curandosi di tralasciare le proprie cure, che di qualche notabile utile li erano, per dare questa sodisfattione, e per comunicare parte della sua profonda dottrina, a quelli che con grandissima attenzione ogni mattina procuravano ascoltarlo."

62. Harvey, *Lectures*, 22.

63. Ibid., 27.

EPILOGUE

1. This sentence comes at the end of a string of descriptions of the true philosopher: "For true Philosophers, who are perfectly in love with truth and wisdom, never find themselves so wise, or full of wisdom, or so abundantly satisfied in their own knowledge, but that they give place to truth whensoever, or from whosoever it comes." See Harvey, *Anatomical Exercises*, xi.

2. Ibid.

3. See Park, *Secrets of Women*.

4. See *Harvey e Padova*. It is well known that Harvey, like his teacher Fabrici, was committed to Aristotelian studies of nature. As Robert Frank has explained, Harvey was to be influenced later by Oxonians such as Nathaniel Highmore, William Petty, Ralph Bathurst, and Thomas Willis, whose ideas about the corpuscular nature of matter, and about solids and fluids, posed serious challenges to the notion of vital spirits that structured Galenic physiology and to the primacy of the heart, those features of Harvey's work that are closely aligned with Aristotelian studies of nature and with his Paduan education. In addition to Frank's *Harvey and the Oxford Physiologists*, see Frank, "Medicine." On the technology of the pump, see Webster, "William Harvey's Conception." Harvey's use of the pump to describe the motions of the heart has also been connected to the system of locks in Padua. See Boyle, "Harvey in the Sluice."

# Bibliography

## MANUSCRIPT SOURCES

Abriano, F. *Annali di padova* (a manuscript collection of transcribed chronicles of Padua), ms. BP 149, Library of the Civic Museum, Padua (MCP)

Cagnoni, G. "I teatri anatomici dell'Università di Padova," tesi di laurea, Università Iuav di Venezia (unpublished undergraduate thesis)

*Epistolario della nazione artisti*, 1565–1647, n. 476–77, Epistolari tedeschi, Archivio Antico dell'Università di Padova, Padua (AAUP)

*Lettere dei riformatori dello studio*, 1555–1559, filza 63, Archivio di Stato, Venice (ASV)

Malfatti, Cesare. *Descrizione particolare della città di Padova et del territorio padoano* (1606), BP 1352.II, MCP

*Padova, Studio. Documenti 1467–1625* (Cod. Cic. 2525/52, n. 22), Library of the Correr Museum, Venice

*Raccolta Minato*, v. 20, AAUP

Rossi, Niccolò. *Annali di Padova* (1562–1620), ms. BP 147, MCP

*Rotoli*, filza 651, f. 173v, AAUP

*Senato Terra*, Registro 66, 12 September 1596, ASV

## PRINTED SOURCES

Agrimi, Jole and Chiara Crisciani. "Per una ricerca su *experimentum-experimenta*: Riflessione epistemologica e tradizione medica (secoli XIII–XV)," in *Presenza del lessico greco e latino nelle lingue contemporanee*, ed. Pietro Janni and Innocenzo Mazzini (Macerata: Università degli Studi di Macerata, 1990), 9–49.

———. *Edocere medicos: medicina scholastica nei secoli XIII–XV* (Italy: Guerini and Associates, 1988).

*Di alcuni trattatelli da codici della Vaticana*, trans. Felice Lombardi (Pisa: Giardini, 1964).

Andrews, Richard. *Scripts and Scenarios: The Performance of Comedy in Renaissance Italy* (Cambridge: Cambridge University Press, 1993).

Ariès, Philippe. *The Hour of Our Death* (New York: Knopf, 1981).

Aristotle, *Complete Works* (Princeton, NJ: Princeton University Press, 1984).

*Atti della nazione germanica artista* [*Acta germanicae artistarum*], ed. Antonio Favaro, 2 vols. (Padua: Typografia Emiliana, 1911–1912).

Bakhtin, Mikhail. *Rabelais and His World* (1965), trans. Helene Iswolsky (Bloomington: Indiana University Press, 1984).

Baldini, Artemio Enzo. "Per la biografia di Francesco Piccolomini," *Rinascimento* 20 (1980): 389–420.

Bareggi, Claudia di Filippo. *Il mestiere di scrivere: lavoro intellettuale e mercato librario a Venezia nel cinquecento* (Rome: Bulzoni, 1988).

Bartolomeo, Franzina. *Collaudatio mortuorum quorum anatomen publice professus est Patavii perillustris et excellentissimus Iulius Casserius Placentinus, Doctor Medicinae, Eques Divi Marci et Chirurgiae in celeberrimo Gymnasio Patavino Professor Publicus, dum funus honorificentissime fieret* (Padua: Laurentii Pasquati, 1614).

Becher, Tony. *Academic Tribes and Territories: Intellectual Enquiry and the Cultures of Disciplines* (Bristol, UK: Open University Press, 1989).

Benedetti, Alessandro. *Anatomice, sive historia corporis humani,* ed. and trans. Giovanna Ferrari (Florence: Giunta, 1998).

———. *Historia corporis humani sive anatomica* (Venice: Bernardino Guerraldo Vercellensis, 1502).

Benetti, Francesca Zen. "La libreria di Girolamo Fabrici D'Acquapendente," in *Quaderni per la Storia dell'Università di Padova* 9–10 (1976): 161–83.

Berengario da Carpi, Jacopo. *Commentaria cum amplissimis additionibus super anatomiam Mundini una cum textu eiusdem in pristinum et verum nitorem redacto* (Bologna: Hieronymum de Benedictis, 1521).

Bernardi, Francesco. *Prospetto storico-critico: dell'origine, facoltà, diversi stati, progressi, e vicende del Collegio Medico-Chirurgico, e dell'Arte Chirurgica in Venezia* (1797), in *Dalla scienza medica alla pratica dei corpi: fonti e manoscritti marciani per la Storia della Sanità* (Venice: Neri Pozza, 1993).

Black, Robert. "Italian Renaissance Education: Changing Perspectives and Continuing Controversies," *Journal of the History of Ideas* 52, no. 2 (1991): 315–34.

Blancken, Gerrard. *A Catalogue of All the Cheifest* [sic] *Rarities in the Publick Theater and Anatomie-Hall of the University of Leiden* (Leiden: Hubert vander Boxe, 1695).

Bloom, Gina. *Voice in Motion: Staging Gender, Shaping Sound in Early Modern England* (Philadelphia: University of Pennsylvania Press, 2007).

Bonuzzi, Luciano. "Medicina e sanità," in *Storia di Venezia: dalle origini alla caduta della Serenissima,* vol. 5, *Il Rinascimento, società ed economia,* ed. Alberto Tenenti and Ugo Tucci (Rome: Istituto della Enciclopedia Italiana, 1996), 407–40.

Botteri, Inge. *Galateo e galatei: la creanza e l'instituzione* [sic] *della società nella trattistica italiana tra antico regime e stato liberale* (Rome: Bulzoni, 1999).

Boyle, Marjorie O'Rourke. "Harvey in the Sluice: From Hydraulic Engineering to Human Physiology," *History of Technology* 24 (2008): 1–22.

Brugi, Biagio. *Gli scolari dello studio di Padova nel cinquecento (con un'appendice su gli studenti tedeschi e la S. Inquisizione a Padova nella seconda metà del secolo XVI),* 2nd ed., rev. (Padua and Verona: Drücker, 1905).

Brunelli, Bruno. "Due accademie padovane del cinquecento," *Atti e Memorie della R. Accademia di Scienze, Lettere ed Arti in Padova,* n. s., 36, disp. 1 (1920): 43–57.

———. "Francesco Portenari e le cantate delgi accademici padovani," *Atti del Reale Istituto Veneto di Scienze, Lettere ed Arti, Parte 2*, 79 (1919–1920): 595–607.

Burckhardt, Jacob. *The Civilization of the Renaissance in Italy*, trans. S. G. C. Middlemore (1937), selected and explained by Ludwig Goldscheider (London: Phaidon Press, 1960).

Burke, Peter. "Early Modern Venice as a Center of Information and Communication," in *Venice Reconsidered: The History and Civilization of an Italian City-State, 1297–1797*, ed. John Martin and Dennis Romano (Baltimore: Johns Hopkins University Press, 2000), 389–419.

———. *Popular Culture in Early Modern Europe* (New York: New York University Press, 1978).

Bylebyl, Jerome. "Interpreting the *Fasciculo* Anatomy Scene," *Journal of the History of Medicine and Allied Sciences* 45 (1990): 285–316.

———. "Medicine, Philosophy, and Humanism in Renaissance Italy," in *Science and the Arts in the Renaissance*, ed. John Shirley and F. David Hoeniger (Washington, DC: Folger Library, 1985), 27–49.

———. "The School of Padua: Humanistic Medicine in the Sixteenth Century," in *Health, Medicine, and Mortality in the Sixteenth Century*, ed. Charles Webster (Cambridge: Cambridge University Press, 1979), 335–70.

Bynum, William. "The Anatomical Method, Natural Theology, and the Functions of the Brain," *Isis* 64, no. 4 (1973): 445–68.

Campion, Edmund. *Rationes decem: quibus fretus, certamen aduersarijs obtulit in causa fidei* (n.l.: Cosmopoli, 1581).

Camporesi, Piero. *The Anatomy of the Senses: Natural Symbols in Medieval and Early Modern Italy*, trans. Allan Cameron (Cambridge: Polity Press, 1994).

Capivacci, Girolamo. *De methodo anatomica* (Venice: Baptistam Ciottum Senensem, 1593).

Carlino, Andrea. "Les fondements humanistes de la médecine: rhétorique et anatomie à Padoue vers 1540," *Littérature et médecine: approches et perspectives (1500–1900)*, ed. Andrea Carlino and Alexandre Wenger (Geneva: Droz, 2007), 19–47.

———. "Petrarch and the Early Modern Critics of Medicine," *Journal of Medieval and Early Modern Studies* 35 (2005): 559–82.

———. *Books of the Body: Anatomical Ritual and Renaissance Learning* (Chicago: University of Chicago Press, 1999), originally published as *La fabbrica del corpo: libri e dissezione nel Rinascimento* (Turin: Einaudi, 1994).

Carruthers, Mary. *The Book of Memory: A Study of Memory in Medieval Culture* (Cambridge: Cambridge University Press, 1990).

Casseri, Giulio. *The Larynx, Organ of Voice*, trans. Malcolm Hast and Erling Holtsmark (Uppsala: Almquist and Wiksells, 1969).

———. *De vocis auditusque organis historia anatomica* (Ferrara: Victorius Baldinus, 1600 [1601]).

Cecchetti, Bartolomeo. *Per la storia della medicina in Venezia* (Venice: Pietro Naratovich, 1886).

Choulant, Ludwig. *History and Bibliography of Anatomic Illustration: In Its Relation to Anatomic Science and the Graphic Arts*, ed. and trans. Mortimer Frank (New York: Schuman, 1945).

Ciappelli, Giovanni. *Carnevale e quaresima: comportamenti sociali e cultura a Firenze nel Rinascimento* (Rome: Storia e Letteratura, 1997).

Cipolla, Carlo M. *Public Health and the Medical Profession in the Renaissance* (Cambridge: Cambridge University Press, 1976).

Clark, William. *Academic Charisma and the Origins of the Research University* (Chicago: University of Chicago Press, 2006).

Cohen, I. Bernard. *Revolution in Science* (Cambridge, MA: Harvard University Press, 1985).

Colombo, Realdo. *De re anatomica libri XV* (Venice: Bevilacqua, 1559).

Costello, William. *The Scholastic Curriculum at Early Seventeenth-Century Cambridge* (Cambridge, MA: Harvard University Press, 1958).

Cremonini, Cesare. *Le orazioni*, ed. Antonino Poppi (Padua: Antenore, 1998).

———. *Il nascimento di Venetia* (Bergamo: Valerio Ventura e Fratelli, 1617).

Crispolti, Cesare. *Idee dello scolare che versa negli studi* (Perugia: Vincentio Colombara, 1604).

Crombie, Alistair. *Styles of Scientific Thinking in the European Tradition* (London: Duckworth, 1994).

Cunningham, Andrew. *The Anatomical Renaissance: The Resurrection of the Anatomical Projects of the Ancients* (Brookfield, VT: Ashgate, 1997).

———. "Fabricius and the 'Aristotle Project' in Anatomical Teaching and Research at Padua," in *The Medical Renaissance of the Sixteenth Century*, ed. Andrew Wear, Roger K. French, and Iain M. Lonie (Cambridge: Cambridge University Press, 1985), 195–222.

Cunsolo, Elisabetta. "Giulio Casserio e la pubblicazione del *De vocis auditusque organis* tra Padova e Ferrara all'inizio del '600," *MEFRIM: Mélanges de l'École Française de Rome, Italie, et Méditerranée* 120, no. 2 (2008): 385–405.

Dal Piaz, Vittorio. "L'orto botanico e il teatro anatomico di Padova: indagini e contributi," *Quaderni per la Storia dell'Università di Padova* 31 (1998): 63–73.

Dal Piaz, Vittorio and Maurizio Rippa Bonati, "The Design and Form of the Padua 'Horto Medicinale,'" in *The Botanical Garden of Padua 1545–1995*, ed. Alessandro Minelli (Venice: Marsilio, 1995), 33–56.

Daniele, A. "Sperone Speroni, Bernardo Tomitano e l'accademia degli infiammati di Padova," *Filologia Veneta* 2 (1989): 1–53.

Daston, Lorraine. "Attention and the Values of Nature in the Enlightenment," in *The Moral Authority of Nature*, eds. Lorraine Daston and Fernando Vidal (Chicago: University of Chicago Press, 2004), 100–126.

———. "The Moral Economy of Science," *Osiris* 10 (1995): 3–24.

———. "Baconian Facts, Academic Civility, and the Prehistory of Objectivity," in *Rethinking Objectivity*, ed. Allan Megill (Durham: Duke University Press, 1994), 37–63.

Daston, Lorraine and Katharine Park. *Wonders and the Order of Nature, 1150–1750* (New York: Zone Books, 1998).

De Bernardin, Sandro. "La politica culturale della repubblica di Venezia e l'Università di Padova nel XVII secolo," *Studi Veneziani* 16 (1974): 443–502.

Dear, Peter. "Towards a Genealogy of Modern Science," in *The Mindful Hand: Inquiry and Invention from the Late Renaissance to Early Industrialisation*, ed. Lissa Roberts, Simon Schaffer, and Peter Dear (Amsterdam: Koninklijke Nederlandse Akademie van Wetenschappen, 2007), 431–41.

———. *Revolutionizing the Sciences: European Knowledge and Its Ambitions, 1500–1700* (Princeton: Princeton University Press, 2001).

———. "Jesuit Mathematical Science and the Reconstitution of Experience in the Early Seventeenth Century," *Studies in History and Philosophy of Science* 18 (1987): 133–75.

de Sandre Gasparini, Giuseppina. "La confraternità di S. Giovanni Evangelista della Morte in Padova e una 'riforma' ispirata dal vescovo Pietro Barozzi (1502)," in *Miscellanea Gilles Gérard Meersseman*, 2 vols. (Padua: Antenore, 1970), vol. 1, 765–815.

de Vivo, Filippo. *Information and Communication in Venice: Rethinking Early Modern Politics* (Oxford: Oxford University Press, 2007).

———. "Pharmacies as Centres of Communication in Early Modern Venice," in *Renaissance Studies* 21 (2007): 505–21.

della Casa, Giovanni. *Il Galateo*, trans. Konrad Eisenbichler and Kenneth Bartlett (Toronto: Centre for Reformation and Renaissance Studies, 1994).

della Croce, Giovanni Andrea. *Chirurgiae universalis opus absolutum* (Venice: Robertum Meiettum, 1596).

Dooley, Brendan. "Social Control and the Italian Universities: From Renaissance to Illuminismo," *Journal of Modern History* 61 (1989): 205–39.

Dryander, Johann. *Anatomia mundini* (Marburg: C. Egenolphi, 1541).

Eamon, William. *Science and the Secrets of Nature: Books of Secrets in Medieval and Early Modern Culture* (Princeton: Princeton University Press, 1994).

Edgerton, Samuel. *Pictures and Punishment: Art and Criminal Prosecution during the Florentine Renaissance* (Ithaca: Cornell University Press, 1985).

Elias, Norbert. *The Civilizing Process* (1939), trans. Edmund Jephcott, 2 vols. (Oxford: Blackwell, 2000).

*The Embryological Treatises of Hieronymus Fabricius of Aquapendente*, ed. and trans. Howard B. Adelmann (Ithaca: Cornell University Press, 1942).

*Epistolario di Gabriele Falloppia*, ed. Pericle di Pietro (Ferrara: Università degli Studi di Ferrara, 1970).

Evelyn, John. *Memoirs of John Evelyn*, ed. William Bray (London: n.p., 1818).

Fabrici [Fabrizi] d'Acquapendente, Girolamo. *I trattati dell'orecchio, organo dell'udito e Della laringe, organo della voce*, ed. and trans. Luigi Stroppiana (Rome: Arti Grafiche e Cosdidente, 1967).

———. *De venarum ostiolis 1603 of Hieronymus Fabricius of Aquapendente*, facsimile ed., ed. Kenneth Franklin (Springfield, IL: Charles C Thomas, 1933).

———. *Opera omnia anatomica et physiologica*, new ed., ed. Bernardi Albini (Lugduni Batavorum: Johannem van Kerckhem, 1738).

———. *Opera chirurgica, quorum pars prior pentateuchum chirurgicum, posterior operationes chirurgicas* (Padua: Francisci Bolzettae, 1647).

————. *De motu locali animalium* (Padua: Jo. Baptistam de Martinis, 1618).

————. *Pentateuchos Cheirurgicum* (Frankfurt: Nicolai Hoffmanni, 1604).

————. *De formato foetu* (Venice: Franciscus Bolzetta, 1600).

————. *De visione, voce, auditu* (Venice: Franciscum Bolzettam, 1600).

Falloppia, Gabriele. *Observationes anatomicae*, ed. Gabriella Righi Riva and Pericle Di Pietro, 2 vols. (Modena: S. T. E. M. Mucchi, 1964).

Favaro, Antonio. *Galileo Galilei e lo studio di Padova* (Padua: Antenore, 1966).

Favaro, Giuseppe. *Gabrielle Falloppia modenese: studio biografico* (Modena: Immacolata Concezione, 1928).

————. "Girolamo Fabrici d'Acquapendente e la medicina pratica," *Bollettino Storico Italiano dell'Arte Sanitaria* 26 (1927): 3–12.

————. "L'insegnamento anatomico di Girolamo Fabrici d'Acquapendente," in *Monografie storiche sullo studio di Padova, contributo del R. Instituto veneto di scienze, lettere ed arti alla celebrazione del VII centenario della università* (Venice: Ferrari, 1922), 107–36.

Fedalto, Giorgio. "Stranieri a Venezia e a Padova, 1550–1700," in *Storia della cultura veneta*, vol. 4, *Il seicento*, ed. Girolamo Arnaldi and Manlio Pastore Stocchi (Vicenza: Neri Pozza, 1984), pt. 2, 251–79.

Ferrari, Giovanna. *L'esperienza del passato: Alessandro Benedetti filologo e medico umanista* (Florence: Olschki, 1996).

————. "Public Anatomy Lessons and the Carnival: The Anatomy Theatre of Bologna," *Past and Present* 117 (1987): 50–106.

Findlen, Paula. "Sites of Anatomy, Botany, and Natural History," in *Cambridge History of Science: Early Modern*, ed. Katharine Park and Lorraine Daston (Cambridge: Cambridge University Press, 2006), 272–89.

————. "Between Carnival and Lent: The Scientific Revolution at the Margins of Culture," *Configurations* 6, no. 2 (1998): 234–67.

Foucault, Michel. *Discipline and Punish* (1977), trans. Alan Sheridan (New York: Vintage Books, 1995).

Frank, Robert. "Medicine," in *The History of the University of Oxford*, vol. 4, *Seventeenth-Century Oxford*, ed. Nicholas Tyacke (Oxford: Clarendon Press, 1997), 552–53.

————. *Harvey and the Oxford Physiologists: Scientific Ideas and Social Interaction* (Berkeley: University of California Press, 1980).

Freedberg, David. *The Eye of the Lynx: Galileo, His Friends, and the Beginnings of Modern Natural History* (Chicago: University of Chicago Press, 2002).

French, Roger. *Dissection and Vivisection in the European Renaissance* (Aldershot, UK: Ashgate, 1999).

————. *William Harvey's Natural Philosophy* (Cambridge: Cambridge University Press, 1994).

————. "*De Juvamentis Membrorum* and the Reception of Galenic Physiological Anatomy," *Isis* 70 (1979): 96–109.

————. "A Note on the Anatomical *Accessus* of the Middle Ages," *Medical History* 23 (1979): 426–63.

Galen. *On the Usefulness of the Parts of the Body* [*De usu partium*], trans. Margaret Tall-madge May, 2 vols. (Ithaca: Cornell University Press, 1968).

———. *On Anatomical Procedures* [*De anatomicis administrationibus*], trans. Charles Singer (New York: Oxford University Press, 1956).

Gamba, Antonio. "Il primo teatro anatomico stabile in Padova non fu quello di Fabrici d'Acquapendente," in *Atti e Memorie dell'Accademia Patavina di Scienze, Lettere ed Arti, Parte 3*, 99 (1986–1987): 157–61.

Garin, Eugenio. *Il pensiero pedagogico dello Umanesimo* (Florence: Giuntine, 1958).

Gaukroger, Stephen. *The Emergence of a Scientific Culture* (Oxford: Oxford University Press, 2006).

Goldthwaite, Richard. *The Economy of Renaissance Florence* (Baltimore: Johns Hopkins University Press, 2009).

Grafton, Anthony and Lisa Jardine. *From Humanism to the Humanities: Education and the Liberal Arts in Fifteenth- and Sixteenth-Century Europe* (Cambridge, MA: Harvard University Press, 1986).

Grandi, Giacomo. *Orazione per il publico anatomico nell'aprirsi il nuovo teatro di anatomia in Venezia* (Venice: Andrea Giuliani, 1671).

Green, Monica. "From 'Diseases of Women' to 'Secrets of Women': The Transformation of Gynecological Literature in the Later Middle Ages," *Journal of Medieval and Early Modern Studies* 30 (2000): 5–39.

Grendler, Paul. "The Universities of the Renaissance and Reformation," *Renaissance Quarterly* 57 (2004): 1–42.

———. *The Universities of the Italian Renaissance* (Baltimore: Johns Hopkins University Press, 2002).

Guazzo, Stefano. *La civil conversazione*, ed. Amedeo Quondam, 2 vols. (Modena: Panini, 1993).

Hacking, Ian. "'Style' for Historians and Philosophers," chapter 12 in *Historical Ontology* (Cambridge, MA: Harvard University Press, 2002).

*A Hangman's Diary: Being the Journal of Master Franz Schmidt, Public Executioner of Nurem-berg, 1573–1617*, ed. A. Keller, trans. C. Calvert and A. Gruner (Montclair, NJ: Patterson Smith, 1973).

Harkness, Deborah. *The Jewel House: Elizabethan London and the Scientific Revolution* (Princeton: Princeton University Press, 2007).

Harvey, William. *The Anatomical Exercises:* De motu cordis *and* De circulatione sangui-nis, ed. Geoffrey Keynes (New York: Dover Publications, 1995).

———. *Lectures on the Whole of Anatomy: An Annotated Translation of* Prelectiones anato-miae universalis, trans. Charles D. O'Malley, F. N. L. Poynter, and K. F. Russell (Berke-ley: University of California Press, 1961).

*Harvey e Padova: atti del convegno celebrativo del quarto centenario della laurea di William Harvey, Padova, 21–22 novembre 2002*, ed. Giuseppe Ongaro, Maurizio Rippa Bonati, and Gaetano Thiene (Treviso: Antilia, 2006).

Haugen, Kristine Louise. "Academic Charisma and the Old Regime," *History of Universi-ties* 22, no. 1 (2007): 199–228.

Heckscher, William. *Rembrandt's* Anatomy of Dr. Tulp: *An Iconological Study* (New York: New York University Press, 1956).

Henderson, John. *The Renaissance Hospital: Healing the Body and Healing the Soul* (New Haven: Yale University Press, 2006).

Heseler, Baldassar. *Andreas Vesalius' First Public Anatomy at Bologna: An Eyewitness Report by Baldassar Heseler,* ed. and trans. Ruben Ericksson (Uppsala: Almquist and Wiksells, 1959).

*Historia: Empiricism and Erudition in Early Modern Europe,* ed. Gianna Pomata and Nancy Siraisi (Cambridge, MA: MIT Press, 2005).

*History from Crime,* ed. Edward Muir and Guido Ruggiero, trans. Corrada Biazzo Curry, Margaret A. Gallucci, and Mary M. Gallucci (Baltimore: Johns Hopkins University Press, 1994).

*The History of Medical Education,* ed. Charles O'Malley (Berkeley: University of California Press, 1970).

*A History of the University in Europe,* vol. 2, *Universities in Early Modern Europe, 1500–1800,* ed. Hilde de Ridder-Symoens (Cambridge: Cambridge University Press, 1996).

*A History of the University in Europe,* vol. 1, *Universities in the Middle Ages,* ed. Hilde de Ridder-Symoens (Cambridge: Cambridge University Press, 1992).

Holz, G. "Le style nu des relations de voyages," in *Le lexique métalittéraire français, XVI–XVII siècles,* ed. Michel Jourde and Jean-Charles Monferran (Paris: Droz, 2006), 165–85.

Horodowich, Elizabeth. *Language and Statecraft in Early Modern Venice* (New York: Cambridge University Press, 2008).

Howell, Wilbur Samuel. *Poetics, Rhetoric, and Logic: Studies in the Basic Disciplines of Criticism* (Ithaca: Cornell University Press, 1975).

Jardine, Nicholas. "Keeping Order in the School of Padua: Jacopo Zabarella and Francesco Piccolimini on the Offices of Philosophy," in *Method and Order in Renaissance Philosophy of Nature: The Aristotle Commentary Tradition,* ed. Daniel Di Liscia, Eckhard Kessler, and Charlotte Methuen (Aldershot, UK: Ashgate, 1997), 195–209.

Kagan, Richard. "Universities in Italy, 1500–1700," in *Les universités européennes du XVIè au XVIIIè siècle: histoire sociale des populations étudiantes,* vol. 1, *Bohême, Espagne, États italiens, Pays germaniques, Pologne, Provinces-Unies,* ed. Dominique Julia, Jacques Revel, and Roger Chartier (Paris: Éditions de l'École des Hautes Études en Sciences Sociales, 1986), 153–86.

Kaiser, David. "Focus—the Generalist Vision in the History of Science," *Isis* 96, no. 2 (2005): 224–51.

Karras, Ruth Mazo. *From Boys to Men: Formations of Masculinity in Late Medieval Europe* (Philadelphia: University of Pennsylvania Press, 2003).

Kemp, Martin. "Temples of the Body and Temples of the Cosmos: Vision and Visualization in the Vesalian and Copernican Revolutions," in *Picturing Knowledge: Historical and Philosophical Problems Concerning the Use of Art in Science,* ed. Brian S. Baigrie (Toronto: University of Toronto Press, 1996), 40–85.

———. "The Mark of Truth: Looking and Learning in Some Anatomical Illustrations from the Renaissance and Eighteenth Century," in *Medicine and the Five Senses,* ed.

William F. Bynum and Roy Porter (Cambridge: Cambridge University Press, 1993), 85–121.

Ketham, Johann de. *Fasciculo di Medicina, Venice 1493* [Latin original 1491, Italian trans. by Sebastiano Manilio 1493], ed. and trans. Charles Singer [English trans. of 1493 version], 2 vols. (Florence: R. Lier, 1925).

Kibre, Pearl. *The Nations in the Mediaeval Universities* (Cambridge, MA: Mediaeval Academy of America, 1948).

Klestinec, Cynthia. "Civility, Comportment, and the Anatomy Theater: Girolamo Fabrici and His Medical Students in Renaissance Padua," *Renaissance Quarterly* 60 (2007): 434–63.

———."A History of Anatomy Theaters in Sixteenth-Century Padua," *Journal of the History of Medicine and Allied Sciences* 59 (2004): 375–412.

Kusukawa, Sachiko. "The Uses of Pictures in the Formation of Learned Knowledge: The Cases of Leonard Fuchs and Andreas Vesalius," in *Transmitting Knowledge: Words, Images, and Instruments in Early Modern Europe*, ed. Sachiko Kusukawa and Ian Maclean (New York: Oxford University Press, 2006), 73–134.

Lambert, Samuel W., Willy Wiegand, and William M. Ivins, Jr. *Three Vesalian Essays to Accompany the* Icones anatomicae *of 1934* (New York: Macmillan, 1952).

Lanfranco. *Chirurgia parva Lanfranci*, trans. John Hall (London: Thomas Marshe, 1565).

Lazzerini, Luigi. "Le radici folkloriche dell'anatomia: scienza e rituale all'inizio dell'età moderna," *Quaderni Storici* 85 (1994): 193–233.

Lind, Levi R. *Pre-Vesalian Anatomy: Biography, Translations, Documents* (Philadelphia: American Philosophical Society, 1975).

Lohr, Charles. "Metaphysics and Natural Philosophy as Sciences: The Catholic and Protestant Views in the Sixteenth and Seventeenth Centuries," in *Philosophy in the Sixteenth and Seventeenth Centuries*, ed. Constance Blackwell and Sachiko Kusukawa (Aldershot, UK: Ashgate, 1999), 280–95.

Lombardelli, Orazio. *Gl'aphorismi scholastici* (Siena: Appresso Salvestro Marchetti, 1603).

Long, Pamela O. *Openness, Secrecy, Authorship: Technical Arts and the Culture of Knowledge from Antiquity to the Renaissance* (Baltimore: Johns Hopkins University Press, 2001).

Lowry, Martin. "The Proving Ground: Venetian Academies of the Fifteenth and Sixteenth Centuries," in *The Fairest Flower: The Emergence of Linguistic National Consciousness in Renaissance Europe*, ed. C. D. Lanham (Florence: Presso l'Accademia [della Crusca], 1985), 41–51.

Maclean, Ian. "White Crows, Graying Hair, and Eyelashes: Problems for Natural Historians in the Reception of Aristotelian Logic and Biology from Pompanazzi to Bacon," in Historia: *Empiricism and Erudition in Early Modern Europe*, ed. Gianna Pomata and Nancy Siraisi (Cambridge, MA: MIT Press, 2005), 147–80.

Magni, Pietro Paolo. *Il modo di sanguinare* (Rome: Barolomeo Bonfadino e Tito Diani, 1584).

Mandressi, Rafael. *Le regard de l'anatomiste: dissections et invention du corps en Occident* (Paris: Seuill, 2003).

Manutio, Aldo. *Eleganze insieme con la copia della lingua toscana, e latina; scielte da Aldo Manutio. Utilissime al comporre nell'una, e l'altra Lingua. Con tre nuove tavole. La prima, de'Capi, la seconda, delle locutioni, la terza, delle locutioni latine* (Venice: Alessandro Griffio, 1585).

Martínez-Vidal, Àlvar and José Pardo-Tomás. "Anatomical Theatres and the Teaching of Anatomy in Early Modern Spain," *Medical History* 49 (2005): 251–80.

Massa, Niccolò. *Liber introductorius anathomiae* (Venice: Jordani Zilleti, 1539 [the colophon states 1536]).

Masten, Jeffrey. "Is the Fundament a Grave?" in *The Body in Parts: Fantasies of Corporeality in Early Modern Europe*, ed. David Hillman and Carla Mazzio (New York: Routledge, 1997), 129–46.

Mazzio, Carla. "The Senses Divided: Organs, Objects, and Media in Early Modern England," in *Empire of the Senses: The Sensual Culture Reader*, ed. David Howes (New York: Berg, 2005), 85–105.

McClure, George. *The Culture of Profession in Late Renaissance Italy* (Toronto: University of Toronto Press, 2004).

McVaugh, Michael. *The Rational Surgery of the Middle Ages* (Florence: Edizioni del Galluzzo, 2006).

*The Medical Renaissance of the Sixteenth Century*, ed. Andrew Wear, Roger K. French, and Iain M. Lonie (Cambridge: Cambridge University Press, 1985).

Meduna, Bartolomeo. *Lo scolare* (Venice: Pietro Fachinetti, 1588).

Molmenti, Pompeo. *La storia di Venezia nella vita privata dalle origini alla caduta della repubblica* (Trieste: Edizioni Lint, 1973).

Mondino. *Anothomia di Mondino de'Liuzzi da Bologna, XIV secolo*, ed. Piero Giorgi and Gian Franco Pasini (Bologna: San Giovanni in Persiceto, 1992).

Mortimer, Ruth. "The Author's Image: Italian Sixteenth-Century Printed Portraits," *Harvard Library Bulletin* 7, no. 2 (1996): 7–87.

Muir, Edward. *Civic Ritual in Renaissance Venice* (Princeton: Princeton University Press, 1981).

Muraro, Michelangelo. "Tiziano e le anatomie di Vesalio," in *Tiziano e Venezia: convegno internazionale di studi (1976)* (Vicenza: Neri Pozza, 1980), 307–16.

Nardi, Luigi and Cesare Musatti, "Dell'anatomia in Venezia," in *Storia medica veneta* (Venice: n.p., 1897).

Nutton, Vivian. "The Rise of Medical Humanism: Ferrara, 1464–1555," *Renaissance Studies* 11 (1997): 2–19.

———. "Humanist Surgery," *The Medical Renaissance of the Sixteenth Century*, ed. Andrew Wear, Roger K. French, and Iain M. Lonie (Cambridge: Cambridge University Press, 1985), 75–99.

O'Malley, Charles. *Andreas Vesalius of Brussels: 1514–1564* (Berkeley: University of California Press, 1965).

Ongaro, Giuseppe. "La medicina nello studio di Padova e nel Veneto," *Storia della cultura veneta*, vol. 3, *Dal primo quattrocento al Concilio di Trento*, ed. Girolamo Arnaldi and Manlio Pastore Stocchi (Vicenza: Neri Pozza, 1981), 76–134.

Ongaro, Giuseppe and Maurizio Rippa Bonati. "L'Ortopedia di Girolamo Fabrici d'Acquapendene," *Atti e Memorie dell'Accademia Galileiana di Scienze Lettere ed Arti* 115 (2002–2003): 201–23.

Osler, William. *The Evolution of Modern Medicine* (New Haven: Yale University Press, 1923).

Ottosson, Per-Gunnar. *Scholastic Medicine and Philosophy: A Study of Commentaries on Galen's Tegni (ca. 1300–1450)* (Naples: Bibliopolis, 1984).

*Padova: Diari e viaggi*, ed. Giuseppe Toffanin (Milan: Marzorati, 1990).

Pagel, Walter. *William Harvey's Biological Ideas: Selected Aspects and Historical Background* (New York: Hafner, 1967).

Palmer, Richard. "Pharmacy in the Republic of Venice," in *The Medical Renaissance of the Sixteenth Century*, ed. Andrew Wear, Roger K. French, and Iain M. Lonie (Cambridge: Cambridge University Press, 1985), 100–117.

———. *The Studio of Venice and Its Graduates* (Padua: Lint, 1983).

———. "Nicolò Massa, His Family, and His Fortune," *Medical History* 25 (1981): 385–410.

———. "Physicians and Surgeons in Sixteenth-Century Venice," *Medical History* 23 (1979): 451–60.

Park, Katharine. *Secrets of Women: Gender, Generation, and the Origins of Dissection* (New York: Zone Books, 2006).

———. "The Life of the Corpse: Division and Dissection in Late Medieval Europe," *Journal of the History of Medicine and Allied Sciences* 50 (1995): 111–32.

———. "The Criminal and the Saintly Body: Autopsy and Dissection in Renaissance Italy," *Renaissance Quarterly* 47 (1994): 1–33.

———. "The Organic Soul," in *Cambridge History of Renaissance Philosophy*, ed. Charles B. Schmitt, Quentin Skinner, and Eckhard Kessler (Cambridge: Cambridge University Press, 1988), 464–84.

———. *Doctors and Medicine in Early Renaissance Florence* (Princeton: Princeton University Press, 1985).

Pasinetti, C. "Il ponte dell'anatomia a Venezia," *Bolletino dell'Istituto Storico Italiano dell'Arte Sanitaria* 6 (1926).

Patrizi, Elisabetta. *La trattatistica educativa tra Rinascimento e Controriforma: l'idea dello scolare di Cesare Crispolti* (Pisa: Istituti Editoriali e Poligrafici Internazionali, 2005).

Payne, Lynda. *With Words and Knives: Learning Medical Dispassion in Early Modern England* (Burlington, VT: Ashgate, 2007).

*Pedagogy and the Practice of Science: Historical and Contemporary Perspectives*, ed. David Kaiser (Cambridge, MA: MIT Press, 2005).

Persaud, Trivedi V. N. *The History of Anatomy: The Post-Vesalian Years* (Springfield, IL: Charles C Thomas, 1997).

Platter, Felix, *Beloved Son Felix: The Journal of Felix Platter, a Medical Student in Montpellier in the Sixteenth Century*, trans. Sean Jennett (London: Frederick Muller, 1961).

Pomata, Gianna. "Malpighi and the Holy Body: Medical Experts and Miraculous Evidence in Seventeenth-Century Italy," *Renaissance Quarterly* 21 (2007): 568–86.

———. "*Praxis Historialis*: The Uses of *Historia* in Early Modern Medicine," in *Historia: Empiricism and Erudition in Early Modern Europe*, ed. Gianna Pomata and Nancy Siraisi (Cambridge, MA: MIT Press, 2005), 105–46.

Portenari, Angelo. *Della felicità di Padova* (Padua: Pietro Paolo Tozzi, 1623).

*The Premodern Teenager: Youth in Society, 1150–1650*, ed. Konrad Eisenbichler (Toronto: Centre for Reformation and Renaissance Studies, 2002).

Premuda, L. and B. Bertolaso, "La prima sede dell'insegnamento clinico nel mondo: l'ospedale de S. Francesco grande in Padova," *Acta Medicae Historiae Patavina* 7 (1960–1961): 61–92.

Prosperi, Adriano. "Consolation or Condemnation: Debates on Withholding Sacraments," in *The Art of Executing Well: Rituals of Execution in Renaissance Italy*, ed. Nicholas Terpstra (Kirksville, MO: Truman State University Press, 2000), 98–117.

Pullan, Brian. *Rich and Poor in Renaissance Venice: The Social Institutions of a Catholic State, to 1620* (Cambridge, MA: Harvard University Press, 1971).

*Questyonary of Cyrurgyens* (London: Robert Wyer, 1541).

Randall, John Herman, Jr. "The Development of Scientific Method in the School of Padua," *Journal of the History of Ideas* 1 (1940): 177–206.

*Relazioni dei rettori veneti in Terraferma*, vol. 4, *Podestaria e capitanato di Padova* (Milan: Giuffrè, 1975).

*Renaissance Letters: Revelations of a World Reborn*, ed. and trans. Robert J. Clements and Lorna Levant (New York: New York University Press, 1976).

Riccoboni, Antonio. *De gymnasio patavino* (Padua: Franciscum Bolzettam, 1598).

Richardson, Ruth. *Death, Dissection, and the Destitute* (London: Routledge and Kegan Paul, 1987).

Ridder-Symoens, Hilde de. "Management and Resources," in *A History of the University in Europe*, vol. 2, *Universities in Early Modern Europe, 1500–1800*, ed. Hilde de Ridder-Symoens (Cambridge: Cambridge University Press, 1996), 155–209.

Ridgeway, William. *The Origin of Metallic Currency and Weight Standards* (Cambridge: Cambridge University Press, 1892).

Rinaldi, Massimo. "Modèles de vulgarisation dans l'anatomie du XVIè siècle: notes sur la *Contemplatione anatomica* de Prospero Borgarucci (1564)," in *Vulgariser la médecine: du style médical en France et en Italie (XVIè et XVIIè siècles)*, ed. Andrea Carlino and Michel Jeanneret (Geneva: Droz, 2009), 35–54.

———. *Arte sinottica e visualizzazione del sapere nell'anatomia del cinquecento* (Bari: Cacucci, 2008).

Rippa Bonati, Maurizio. "L'anatomia 'teatrale' nelle descrizioni e nell'iconografia," in *Il teatro anatomico: storia e restauri*, ed. Camillo Semenzato (n.l.: Limena, 1994), 74–76.

———. "Le tradizioni relative al teatro anatomico dell'Università di Padova con particolare riguardo al progetto attribuito a fra' Paolo Sarpi," *Acta Medicae Historiae Patavina* 35–36 (1988–1990): 145–68.

Rizzi, G. *Il teatro anatomico di S. Giacomo dell'Orio a Venezia* (Milan: <n.p.?>, 1957).

Rosand, Ellen. "Music and the Myth of Venice," *Renaissance Quarterly* 30 (1977): 511–37.

Rossetti, Lucia. "Le biblioteche delle 'nationes' nello studio di Padova," *Quaderni per la Storia dell'Università di Padova* 2 (1969): 53–67.

Roth, Moritz. *Andreas Vesalius Bruxellensis* (Berlin: G. Reimer, 1892).

Ruggiero, Guido. *The Boundaries of Eros: Sex Crime and Sexuality in Renaissance Venice* (New York: Oxford University Press, 1985).

Rupp, Jan C. "Matters of Life and Death: The Social and Cultural Conditions of the Rise of Anatomical Theaters, with Special Reference to Seventeenth-Century Holland," *History of Science* 28 (1990): 263–87.

Saibante, Mario, Carlo Vivarini, and Gilberto Voghera. "Gli studenti dell'Università di Padova dalla fine del '500 ai nostri giorni (studio statistico)," *Metron* 5 (1924): 163–223.

Sappol, Michael. *A Traffic of Dead Bodies: Anatomy and Embodied Social Identity in Nineteenth-Century America* (Princeton: Princeton University Press, 2002).

Sawday, Jonathan. *The Body Emblazoned: Dissection and the Human Body in Renaissance Culture* (New York: Routledge, 1995).

———. "The Leiden Anatomy Theatre as a Source for Devanant's 'Cabinet of Death' in 'Gondibert,'" *Notes and Queries* 30 (1983): 437–39.

Schmitt, Charles. "Aristotle among the Physicians," in *The Medical Renaissance of the Sixteenth Century*, ed. Andrew Wear, Roger K. French, and Iain M. Lonie (Cambridge: Cambridge University Press, 1985), 1–15.

Schott, Francesco. *Itinerario, overo nova descrittione de'viaggi principali d'Italia* (Padua: Cadorin, 1659).

Shapin, Steven. *A Social History of Truth: Civility and Science in Seventeenth-Century England* (Chicago: University of Chicago Press, 1994).

Shapin, Steven and Simon Schaffer. *Leviathan and the Air-Pump: Hobbes, Boyle, and the Experimental Life* (Princeton: Princeton University Press, 1985).

Shapiro, Barbara. *The Culture of Fact: England, 1550–1720* (Ithaca: Cornell University Press, 2000).

Shapiro, Meyer. "Style," in *Anthropology Today: An Encyclopedic Inventory*, ed. Alfred L. Kroeber (Chicago: University of Chicago Press, 1953), 287–312.

Siraisi, Nancy. *History, Medicine, and the Traditions of Renaissance Learning* (Ann Arbor: University of Michigan Press, 2007).

———. "*Historia, actio, utilias*: Fabrici e le scienze della vita nel cinquecento," in *Il teatro dei corpi: le pitture colorate d'anatomia de Girolamo Fabrici d'Acquapendente*, eds. Maurizio Rippa Bonati and José Pardo-Tomás (Milan: Mediamed, 2004), 63–73.

———. "History, Antiquarianism, and Medicine: The Case of Girolamo Mercuriale," *Journal of the History of Ideas* 64 (2003): 231–51.

———. "Vesalius and Human Diversity in 'De humani corporis fabrica'," *Journal of the Warburg and Courtauld Institutes* 57 (1994): 60–88.

———. *Medieval and Early Renaissance Medicine: An Introduction to Knowledge and Practice* (Chicago: University of Chicago Press, 1990).

———. *Avicenna in Renaissance Italy: The Canon and Medical Teaching in Italian Universities after 1500* (Princeton: Princeton University Press, 1987).

———. *Taddeo Alderotti and His Pupils: Two Generations of Italian Medical Learning* (Princeton: Princeton University Press, 1981).

Smith, Pamela. *The Body of the Artisan: Art and Experience in the Scientific Revolution* (Chicago: University of Chicago Press, 2004).

*Statuta almae universitatis d. artistarum et medicorum patavini gymnasii* (Padua: Innocentium Ulmum, 1570).

*Statuti del comune di Padova*, trans. Guido Beltrame, Guerrino Citton, and Daniela Mazzon (Padua: Biblos, 2000).

Stella, Aldo. "Intorno al medico padovano Nicolò Buccella, anabaptista del '500," *Atti e Memorie dell'Accademia Patavina di Scienze, Lettere ed Arti, Parte 3*, 74 (1961–1962): 333–61.

Sterzi, G. "Guilio Casserio, anatomico e chirurgo (c. 1552–1616)," *Nuovo Archivio Veneto*, ser. 3, 18 (1909): 207–78 [pt. 1]; 19 (1910): 25–111 [pt. 2].

Strocchia, Sharon. *Death and Ritual in Renaissance Florence* (Baltimore: Johns Hopkins University Press, 1992).

Stroup, Alice. *A Company of Scientists: Botany, Patronage, and Community at the Seventeenth-Century Parisian Royal Academy of Sciences* (Berkeley: University of California Press, 1990).

Sugg, Richard. *Murder after Death: Literature and Anatomy in Early Modern England* (Ithaca: Cornell University Press, 2007).

*Il teatro dei corpi: le pitture colorate d'anatomia di Girolamo Fabrici d'Acquapendente*, ed. Maurizio Rippa Bonati and José Pardo-Tomás (Milan: Mediamed, 2004).

Temkin, Oswei. *Galenism: Rise and Decline of Medical Philosophy* (Ithaca: Cornell University Press, 1973).

Terpstra, Nicholas. "Theory into Practice: Executions, Comforting, and Comforters," in *The Art of Executing Well: Rituals of Execution in Renaissance Italy*, ed. Nicholas Terpstra (Kirksville, MO: Truman State University Press, 2000), 118–158.

Tomasini, Jacopo. *Gymnasium patavinum* (Udine: Nicholai Schriatti, 1654).

Tosoni, Pietro. *Della anatomia degli antichi e della scuola anatomica padovana* (Padua: Del Seminario, 1844).

*Trattati di poetica e retorica del cinquecento*, ed. Bernard Weinberg (Bari: G. Laterza, 1970).

Traub, Valerie. "Mapping the Global Body," in *Early Modern Visual Culture: Representation, Race, and Empire in Renaissance England*, ed. Peter Erickson and Clark Hulse (Philadelphia: University of Pennsylvania Press, 2000), 44–97.

Underwood, E. Ashworth. "The Early Teaching of Anatomy at Padua, with Special Reference to a Model of the Padua Anatomical Theatre," *Annals of Science* 19, no. 1 (1963): 1–26.

*Venice, Città Excelentissima: Selections from the Renaissance Diaries of Marin Sanudo*, ed. Patricia Labalme and Laura White, trans. Linda Carroll (Baltimore: Johns Hopkins University Press, 2008).

Vesalius, Andreas. *On the Fabric of the Human Body*, ed. and trans. William F. Richardson and John B. Carman, vols. 1–3 (San Francisco: Norman Publishing, 2002).

———. *De humani corporis fabrica* (Basel: Oporinus, 1543).

Veslingus, Johann. *Syntagma anatomicum* (Padua: Pauli Frambotti, 1647).

Vicary, Thomas. *The Englishe Mans Treasure, or Treasor for Englishmen: With the true anatomye of mans body* (London: John Windet, 1586).

*Vulgariser la médecine: du style médical en France et en Italie, XVIè et XVIIè siècles*, ed. Andrea Carlino and Michel Jeanneret (Geneva: Droz, 2009).

Webster, Charles. "William Harvey's Conception of the Heart as a Pump," *Bulletin of the History of Medicine* 39 (1965): 508–17.

Weinberg, Bernard. *A History of Literary Criticism in the Italian Renaissance* (Chicago: University of Chicago Press, 1961).

West, William N. *Theatres and Encyclopedias in Early Modern Europe* (Cambridge: Cambridge University Press, 2002).

Yates, Francis. *The Art of Memory* (London: Routledge, 1966).

Zerbi, Gabriele and Maximianus Etruscus. *Gabriele Zerbi, Gerontocomia: On the Care of the Aged; and Maximianus, Elegies on Old Age and Love,* trans. Levi R. Lind (Philadelphia: American Philosophical Society, 1988).

# Index

administrators, 1, 14, 43, 73, 100, 137
aesthetics, xvi, 6, 17, 146
Alderotti, Taddeo, 192n13
Aldrovandi, Ulisse, 66
Aloisius, Michael, 70
anatomists: authority of, 29, 32; and
  Benedetti, 28; and cadavers, 105; and
  commentary tradition, 29; as *demonstrator*
  *(ostensor)*, 23, 31, 32, 34, 42, 43; dominance
  of setting by, 18; and executioner, 130, 131;
  and Falloppio, 49–50; humanist inclinations of, 26, 28; as *lector*, 23, 28, 43; lectures
  by, 24; Massa on, 30; orchestration of public
  demonstration by, 23–24; and pay, xii, 79;
  as philosophers, 63; and private demonstrations, 40; reflectiveness as teachers, 11;
  revelation of secrets by, 18; and rhetoric, 30;
  and second permanent theater, 97, 98,
  100, 106; and state power, 130; students as,
  83, 84, 100, 132, 150, 162; and Vesalius, 32,
  35, 43
anatomy, 23, 31, 61, 71, 146, 149; abnormal,
  22; as basis for universal knowledge, 58;
  and Casseri, 159; civic importance of, 131;
  cultural significance of, 121; of different
  species, 58, 60; and Fabrici, 51, 109, 110;
  and Galen, 44; general, 21; and Harvey, 165;
  and medicine and surgery, xv; and natural
  philosophy, 5, 12–13, 58–62; normative,
  22, 102; refined public tradition of, 126; as
  specialized inquiry, 54. *See also* morphology; structures; use
Andernacus, Johannes, 38
Apellato, 42, 43
apothecaries, 91. *See also* pharmacies
Argent, D., 167, 168

Aristotle, 2, 7, 39, 194n40; and animal and
  human physiology, 210n38; and Capivacci,
  63–64, 65; and Cremonini, 108–9; and
  della Casa, 115; and Fabrici, xix, 13, 51–52, 56,
  60, 61, 62, 65, 101, 102; and Falloppio, 46;
  and Harvey, 231n4; *History of Animals*, 61;
  and humanism, 108, 109; and *intellectus*
  *speculativus* and *intellectus practicus*, 58;
  *On the Generation of Animals*, 61; *On the*
  *Parts of Animals*, 61, 133; *On the Soul*, 61;
  *Poetics*, 108, 109; *Rhetoric*, 108, 113, 213n70
Atanagi, Dionigi, 109
audiences, xii, 18, 21, 53; academic, 24; as on
  display, 78; disruptions by, 90–92, 93;
  exchanges with, 31; and Fabrici, 8, 104,
  120–21, 122; and Falloppio, 46–47; and first
  permanent theater, 73, 78; listening skills
  of, xvi; nonacademic, xix, 14, 24, 126, 145,
  147–48; and private demonstrations, 13, 40,
  126; and public demonstrations, 13, 14, 30,
  41, 126; and seating, 8, 28, 77, 79, 93, 100;
  and second permanent theater, 98, 100,
  102–5, 120–21, 122, 147, 148; size of, xvi, 6;
  student control of, 19; student exclusion of,
  126–27; of temporary theaters, 28; in
  Venice, 31
auditory experiences: and demonstrations, 94;
  and Fabrici, 6, 97; and public demonstrations, xvi, xix, 5, 28; and second permanent
  theater, 97, 100, 106, 123
autopsies, 14, 22–23, 31, 179n37
Avicenna, *Canon*, 12, 20, 30, 194n39

barbers, 3, 18, 34, 64
barber-surgeons, 18, 34, 148
Bembo, Bernardo, 30